Configuration Management

Implementation, Principles, and Applications for Manufacturing Industries

Configuration Management

Implementation, Principles, and Applications for Manufacturing Industries

Joseph Sorrentino

President/CEO, Lean Quality Systems
Dana Point, California

CRC Press
Taylor & Francis Group
Boca Raton   London   New York

CRC Press is an imprint of the
Taylor & Francis Group, an **informa** business

CRC Press
Taylor & Francis Group
6000 Broken Sound Parkway NW, Suite 300
Boca Raton, FL 33487-2742

© 2009 by Taylor & Francis Group, LLC
CRC Press is an imprint of Taylor & Francis Group, an Informa business

International Standard Book Number-13: 978-1-4200-7512-0 (Hardcover)

Library of Congress Cataloging-in-Publication Data

Sorrentino, Joseph N.
　　Configuration management : implementation, principles, and
　　applications for manufacturing industries / Joseph Sorrentino.
　　　　p. cm.
　　Includes index.
　　"A CRC title."
　　ISBN 978-1-4200-7512-0 (alk. paper)
　　1. Project management. 2. Configuration management. I. Title.

TA190.S6685 2009
658.4'04--dc22
                                                                        2008032892

Visit the Taylor & Francis Web site at
http://www.taylorandfrancis.com

and the CRC Press Web site at
http://www.crcpress.com

Contents

Foreword vii

Acknowledgments ix

About the author xi

Introduction: new management enterprise xiii

One   Turning perception into reality with configuration
               management 1

Two   Configuration management: from concept to
               completion using management by objective evidence 3

Three   Contractual requirements 7

Four   The "lost tribal wisdom" of project management 11

Five   Measuring instruments and the "lost art of the hand
               tool" 73

Six   Measuring and analysis for continuous improvement 93

Index   97

Foreword

I would like to take this opportunity to let you know why I decided to write this book. The reasons are mixed with personal history, and a desire to share and mentor those who are still learning how to become managers for the 21st century with the same generosity that has been shown to me over the years.

As a young boy, I recall being awakened in the middle of the night by my father's cry for help after he had suffered a stroke. It was in December 1960, and he died five days later on Christmas Day—the death certificate was dated December 26 to soften the blow. Because my mother had cancer, I assumed the role of father figure, taking care of my mother and ten-year-old younger brother. I went to school during the day and worked in a delicatessen in the evening stocking shelves, delivering beer and whiskey in the neighborhood, and booking numbers. As an 18-year-old boy in Jersey City, I was hardened, scared, and not ready to take on the world. However, the choices were limited: either live in Jersey City in survival mode with no formal education, or join the Navy to learn a trade. The choice was easy. I completed my education in the Navy, and after a short attitude adjustment by the Navy I received incredible opportunities to learn, and I excelled in a love for life and the art of receiving and sharing knowledge.

Today, I work with companies in the public and private sectors, in industries ranging from aerospace to manufacturing to the U.S. military, including the Navy, Marines, Air Force, and Army. It saddens me to see the changes that financial motivation and greed have brought to many organizations, with no consideration for the personal growth of the employees. While many companies "talk the talk," few "walk the walk." Does the word *downsize* mean anything to you?

I write this book to share some of the general knowledge that I believe is necessary to set a foundation for personal growth. The information may not entirely fit your needs, but that is not the sole intent of this book. The intent is to enlighten you regarding the sterile corporate insensitivity that has materialized around us over the past 10 years, and give you the gift of knowledge. I want to provide you with a different mind-set and perhaps give you the foundation for a different strategy for understanding how to survive in today's business structure. I believe you will understand that the gift is not just knowledge, but also logic and intelligence with elasticity that is a hidden gift. When you finish reading this book, you should feel like your happiness and destiny are in your hands, and not your employer's. If you recognize this and reevaluate your direction in life, I have done my job for today.

I will leave you with two thoughts:

1. "Life stinks and I'm part of it."
2. "To take a helping hand, I have to give a helping hand."

Think about it.

Sincerely,
Joe Sorrentino

Acknowledgments

I would like to first thank God for his grace and his loving messengers that have protected me and guided me.

I am grateful to my wife, Laurie, for her gifts of love and encouragement that make me feel as if every day is a new beginning. From the day we met, her unconditional love and encouragement have opened the door to a new life, with rewards beyond belief. I love you.

To Suzanne Collier for her sweet and calm technique that is tough enough to keep me focused on the tasks at hand. Her expertise takes my thoughts and writings and expresses them with the grace and swiftness of a gazelle. I thank you.

To my old-time friend and mentor, Benny Beaver. I met Benny in 1967, when I was first assigned as a nondestructive testing inspector in the U.S. Navy. Benny mentored me and gave me a foundation that was built on integrity and truth in inspections. His passed-down values have been my code of ethics.

I would also like to thank Michael Vermette, the smartest Bubblehead I have ever met, with a heart of gold. And Douglas Conner, a quality assurance (QA) man who has stayed focused on quality, despite little support from his peers. Their help and sharing of knowledge have been instrumental in my decision to write this book.

About the author

For nearly three decades, Joseph Sorrentino has been instrumental in implementing successful quality management systems for commercial companies and government agencies throughout the United States. A retired U.S. Navy quality assurance specialist, Sorrentino is certified as a Level III examiner for visual, magnetic particle, dye penetrant, and ultrasonic inspections.

Sorrentino began his career as a quality management professional in the U.S. Navy, initially as chief petty officer, Level III NAVSEA examiner/quality assurance division officer. After spending eight years as a GS1910 quality specialist stationed at Lakehurst Naval Air Engineering in Lakehurst, New Jersey, and the Long Beach Naval Shipyard in Long Beach, California, Sorrentino began consulting in the private sector for military and aerospace contractors, including The Boeing Company, Allied Signal/Honeywell, the U.S. Army Corps of Engineers, and the U.S. Navy.

As president and CEO of Lean Quality Systems, Inc., Sorrentino specializes in implementing new standard methods for the corporate sector and has successfully worked with more than 25 corporations in the southern California area alone. Sorrentino and his team of highly qualified consultants are experts in quickly assessing management systems and processes and recognizing problem areas, bottlenecks, and waste. Sorrentino himself is recognized within the corporate sector as one of the few consultants who not only tells you what needs to be done, but also shows you how to do it.

Sorrentino resides in Dana Point, California, with his wife, Laurie Zagon, a well-known artist and the founder of Art & Creativity for Healing, Inc., of Laguna Niguel, California. In his spare time, Sorrentino enjoys surfing, sailing, and spending time with his grandchildren and his dog Uno.

Introduction: new management enterprise

New management enterprise is based on understanding configuration management. However, what is configuration management? Who is the configuration manager? For the purposes of this book, the term *configuration management* will be used to define both products and services.

Overview of configuration management

1. *Configuration management (CM)*: The sole purpose of configuration management is to ensure that a product maintains the same design, materials, composition, or processing as was originally intended, from delivery through its entire life cycle. If modifications are needed to meet evolving technology requirements, or are necessary for the application of the product or service, the CM must ensure that the changes are integrated with existing systems, and updated on the original drawings and technical documents. This also relates to assigning liability to the user, manufacturer, designer, or concept visionary in the case of disaster or mishap.

2. *How is CM applied?* How much is required? First, I would like to illustrate my point with the following analogy of how a painting is viewed by the artist versus the collector.
   - An abstract artist creates a painting by beginning with a perception of what he or she would like to express through the painting. As soon as the painting process begins, the artist is driven by an emotional vision of the end result. The abstract collector will then view the abstract "masterpiece" and either love it or hate it based on several factors: (1) does the art collector know the artist's style, passion, and motivation? (2) Can the collector feel the emotion of the art? And (3) does the art take the collector on a viewing journey through the masterpiece and gently return the collector to his or her viewing spot?
   - The realist artist, on the other hand, will paint a masterpiece of a group having a picnic in the park. The realist will try to produce

the visual of the park and picnickers, creating a feeling of the mood and/or conditions of the event. The realist collector will view the art in the same way as the abstract collector. (1) Does the art collector know the artist's style, passion, and motivation? (2) Can the collector feel the emotion of the art? And (3) does the art take the collector on a viewing journey through the art and gently return the collector to his or her viewing spot?

- Although the art is extremely diverse in nature, and there are many differences between abstract and realistic art, the reality is that there are collectors who appreciate both. The reason that these collectors add art to their collections is the same—personal satisfaction. So why do collectors pay hundreds of thousands of dollars for art? Because it fills a need of the collector, and although it may not be functional, viewing the art gives a sense of euphoria. In the business community, the reason for purchasing a product is not that different.

- You may be asking what this has to do with configuration management. I use this analogy to demonstrate that configuration management is driven by the designer, and not the end user or visionary. In illustrating this example, we can state that beauty is in the eye of the beholder, or the customer, but that the configuration items of the product must remain consistent through the life of the product in order to maintain the interest of the buyer.

Overview of configuration items

Configuration management is based on the control of configuration items (CI), and CIs are the key to configuration management. If a change is made to the form, fit, or function of a CI, the safety of the product could be breached. As mentioned previously, product is not the only tangible item in CM; services are also part of the CM process. For example, let's define a simple CI for the application of a new product that has not been introduced yet. The product is a combination toaster oven that bakes and broils. Now, think about the end product and the severity of problems that might be experienced in relation to the form, fit, function, or safety of the toaster oven's different components. In Table I.1 is a list of toaster oven items and their severity rating. In Tables I.2 through I.5, I have given examples of four charts covering risk priority numbers and severity, condition definitions, probability descriptions, and severity, safety, and reliability. I developed this guide for one of the companies that I have worked with during my career. If you view and compare the items and severity ratings in Table I.1 to the four descriptive charts in Tables I.2 through I.5, you will see that not only did we develop the severity rating guide for determining the configuration items, but we also added the operating conditions required by the companies.

Table I.1 Toaster Oven Items and Severity Rating

Item	Severity Rating
Heating coil	1A Unacceptable[a]
Plug	3C Undesirable
Electrical cord	1A Unacceptable
Ground connection	1A Unacceptable
Access door	3E Acceptable with review
Shelving	3E Acceptable with review
Outer box	3C Undesirable
Inner compartment	4D Acceptable as is
Insulation	2D Undesirable
Paint	4B Acceptable with review

[a.] See Table I.3 for definitions of these ratings.

Table I.2 Risk Priority Numbers and Severity

Frequency of Occurrence	1 Catastrophic	2 Critical	3 Marginal	4 Negligible
A: Frequent	1A Unacceptable[a]	2A Unacceptable	3A Unacceptable	4A Acceptable with review
B: Probable	1B Unacceptable	2B Unacceptable	3B Undesirable	4B Acceptable with review
C: Occasional	1C Unacceptable	2C Undesirable	3C Undesirable	4C Acceptable as is
D: Remote	1D Undesirable	2D Undesirable	3D Undesirable	4D Acceptable as is
E: Improbable	1E Acceptable with review	2E Acceptable with review	3E Acceptable with review	4E Acceptable as is

[a.] See Table I.3 for definitions of these ratings.

Table I.3 Conditions Definitions

Unacceptable	Safety/reliability advisory needed
Undesirable	Safety/reliability advisory needed
Acceptable with review	Safety/reliability advisory needed
Acceptable as is	No further action required

Table 1.4   Probability Descriptions

Probability	Description
Frequent	Likely to occur frequently
Probable	Will occur several times during operations
Occasional	Likely to occur sometime during an operation, but not likely to occur during each operation
Remote	Unlikely, but possible to occur during an operation
Improbable	So unlikely that one may assume occurrence may not be experienced

Table 1.5   Severity, Safety, and Reliability

Severity	Safety	Reliability
Catastrophic	Death	Loss of facility
Critical	Severe injury	1. Operations stop 2. Major system or equipment damage or loss
Marginal	Minor injury	1. Degraded performance 2. Significant operations delay
Negligible	No injury	No operational significance

As you can see from the information in Tables I.1 through I.5, we have defined the key characteristics of the toaster oven.

The next step is to pick the CIs based on liability, form, fit, function, and safety. One of the key mistakes made in the industry as a whole during this process is to allow engineering to define the CIs. Occasionally, the quality department is part of the process, but only as an afterthought, or after the drawings and specifications have been developed, meaning that the drawings and specifications will have to be reworked at some point during the process.

Many companies outsource the manufacturing of their products to other companies or to other countries. In one case, I was personally involved in a company that delivered a sample of a product to the Far East, and had them develop a replica of the product for shipment back to the United States to be sold.

In cases where products are shipped internationally for production, the loss of configuration management by the company that owns the product and its liability can happen quite often. In this case, the U.S.-based company was responsible for the ownership and liability of the product, even though the manufacturing was outsourced to the Far East. If the product caused a catastrophic event such as fire, severe injury, or death, the U.S.-based company would be liable.

In another example, I was part of an investigation to identify the root cause and corrective/preventive measures of an occurrence of product failure for a company in Chino, California. The company had used a Chinese company to manufacture a product they had designed. The product had been in production for three years with no negative reports from the consumer, but then came reports of a unit fire with property damage caused by one of the products. During my investigation, I asked for the drawings and specifications of the product being produced in China. As the quality management department presented the documents to me, they said that they had the original drawings and specifications from three years ago, but they knew that there had been changes to the product in order to increase the profit margin, and they did not have a copy of any revisions.

All the revisions had been approved by sales and marketing (red flag #1). As I reviewed the drawings, I noted there were no defined specifications for the materials (red flag #2), only a generic description of the materials. I also noted that because there were no reported problems during the first article inspections, new product runs were given to the Chinese manufacturing plant (red flag #3) in order to remove the overhead of a higher paid quality person in Chino. The company had only reviewed documentation of tests and inspections. Most of the documents were in Chinese and English, but were of poor quality (red flag #4) because they were reproductions of reproductions that were faxed to the company before shipment from China. The bottom line is that anyone with an eighth-grade education could have figured this problem out. The bean counters and the bean heads had been driving the company's manufacturing decisions, and the financial analysts and sales and marketing representatives were making decisions that were out of their area of expertise.

The selection of CIs must be made by a team, and cannot be made by one person or department. The decision must be made not only for safety, but also for any attribute that would affect any part of the organization in an adverse way. What sounds like a good idea to save two cents on each material item, multiplied by 500,000 items a year, may seem good today, but your company could be liable for millions of dollars of loss in a single week if the material fails and causes any damage.

With this as a foundation to work from, we will explore a number of areas in the lost, tribal wisdom section, and you will see the configuration items appear time and time again. The trick is to determine the severity of the configuration item and how to apply good common sense to the way you do business.

You are the configuration manager. Regardless of what stage of the operation you are in, configuration management is your business. You will need to expand your view of the organization, because as outsourcing expands your view of a process that you don't know about, it could mean the difference between working and pushing a shopping cart with your belongings in black plastic bags.

Chapter one

Turning perception into reality with configuration management

Configuration management is designed to turn perception into reality. When building a home:

1. The designer will create an artist's rendering of what the home will look like. This is the first stage in capturing the perception of what the home will feel like, and yes, I do mean what the home will feel like. Remember, at this point the home is still a concept.
2. The architectural engineer will then create drawings of the home incorporating the visual concept and his or her perception of what the home will look like, adding the building code specifications. This is the second stage in perception, and a deeper look into the overall perception.
3. The realization of the perception is the builder's understanding of the architectural drawings.

This example of perception to realization illustrates that a visionary provides perception, an educated engineer completes the architecture, and the tradesperson does the building. The educational factor and skill sets are a roller coaster of highs and lows, and the configuration manager does not typically have a say at the beginning of a project; thus, the only tool is his or her ability to initially review the package and identify problems and delays before the start of the work. Prework reviews are described in detail in chapter four of this book.

The biggest problems inherent in every project are the cost overrides, the late delivery of goods and services, and almost always the completion date. To top it off, the house may not turn out the way it was originally intended.

Some common problems that occur when building a house are poor planning and coordination, inaccurate drawings, codes that are not understood, ambiguous statements in the contract, scheduling of key events, and the like. In the past, it was common to have a construction site engineer, a designer available, and a contractor, supervisors, and employees for each trade on the job site.

In today's market, you are lucky if you get an older technician who is an excellent tradesperson in one of the disciplines, such as framing or electrical work. This person may have had years of experience working on job sites, and can read drawings and understand building specifications and codes. Their years of on-the-job training give them the skill set to know what it takes to get the job done.

The competition in today's market is dog-eat-dog. Most companies are one- or two-person operations that will subcontract almost every aspect of the work. They do this so they have low overhead. It is not uncommon for these subcontractors to make staff changes on a daily basis, because their main concern is to get the job done quickly and cheaply. The more money that is saved in material and labor, the more profit for the company.

The job description for a work coordinator a few years ago was simple because the support team was strong and companies had tradespeople readily available. Most of today's small to midsized companies do not have this luxury—they have become lean and mean. I know they are lean, but instead of mean, I believe they have made themselves green—meaning that they have removed their knowledge base. Engineers and managers fresh out of college are not equipped to deal with today's challenges. They are over-trained in theory and undertrained in practical application. And there are no mentors in these organizations with time to guide and nurture their growth in the industry.

Major corporations, companies, the military, and the civil service have created a new kind of manager—the "configuration manager." The ideal configuration manager should have a thorough background in new and old techniques, with a diverse knowledge and wisdom base. The remainder of this book is meant to give examples of some of the "lost tribal knowledge." Each organization must determine the extent of tribal knowledge necessary to stay one step ahead of the subcontractors hired to accomplish the work. The company or organization liable for the work should be driving the bus, and not be a passenger.

Chapter two

Configuration management: from concept to completion using management by objective evidence

All processes, operations, and functions have objectives. These objectives are the basis for monitoring success, and are the organization's first-line stability indicators. It is necessary to not only identify objectives, but also to maintain objective evidence in order to analyze the stability and success of the project, and maintain a history of the project. Objective evidence is also an important factor in determining the stability of your employees.

Quality objectives are established prior to any movement required to accomplish a task, or even start a business. Unlike the organization's quality policy, quality objectives are determined by the tasks that meet the requirements of the quality policy. Policies are established to guide decisions and implementation methods with projects. Objectives are considered to be ambitions, aims, goals, marks, or targets that can only be determined by measurements. When a policy expresses continuous improvement, the next step is to set measurable objectives for the processes and products to determine if the goal can be achieved.

Quality planning must include the identification and determination of quality objectives in system processes.

Quality objectives are established throughout the organization to support the quality policy, to meet requirements for products and processes, and to improve the organization's systems by measuring performance.

Quality objectives define the direction and priorities for continuous improvement.

Quality objectives define the classification of objective evidence that must be captured, and there are four major categories:

- *Policy objectives*: These are principal, strategic objectives that apply to the whole organization. They are typically included in the quality policy itself, or may be communicated in memoranda from the top management. The president/CEO of a company authorizes policy objectives.
- *Quality performance objectives*: These objectives set specific, measurable targets for improving operational performance to ensure product conformity and customer satisfaction. They apply to departments

and functions having direct responsibility for activities that require improvement. Performance objectives are established, documented, and monitored within the framework of management reviews of the organizations. Management review procedures typically define them in the meeting agendas.

- *Product quality objectives*: These objectives pertain to improvement of products and associated services. The president/CEO and top executive managers responsible for marketing and product development establish product objectives. They can be documented in product briefs, memoranda, or minutes of meetings, and they apply to functions responsible for the research, design, and development of products and services.
- *Quality system objectives*: These objectives pertain to the improvement of quality system processes and performance. Quality system objectives are established, documented, and monitored within the framework of management reviews of the quality system, and in accordance with management review procedures.

Quality system planning

Quality system planning is designed to correctly capture data for which objectives, quality system elements, and processes are planned, while ensuring the effectiveness and efficiency of the quality system in meeting its intended purpose. The purpose of the quality system is as follows:

- To achieve the quality policy
- To ensure and demonstrate the organization's ability to consistently provide customer and regulatory requirements
- To ensure a high level of customer satisfaction
- To facilitate continuous improvement
- To comply with requirements of the organizational operating system

The output of quality system planning is documented in a policy manual, in associated operational procedures, and in other referenced documents. These documents identify and define all elements and processes of the quality system. The objectives are captured in the key elements of the system.

To give an example of how objective evidence is captured and reported, Figure 2.1 shows a monthly cost-of-poor-quality (COPQ) report. This report illustrates, in dollars and cents, the effects of poor quality and monthly losses. This is easy to understand and gives a clearer picture of the loss. If data are the only source of information, and exclude man-hours and material losses, the data information is useless.

In the COPQ at the end of this chapter for the shop nonconforming material report (NCMR), there was only one discrepancy, but the waste in man-hours and material exceeded $50,000, and the internal audit found ten discrepancies, costing the organization a little more than $1,000. This poses the question as to whether one of the internal audit findings caused the shop's NCMR. This is where the fun is—examining data for root cause and effects.

Obviously we could fire the shop supervisor if we went only by the data and amount of loss, or we can be smarter and review the internal corrective action reports (CAR) for indications of a system that is out of control. My first action would be to ask the five most important questions: why, why, why, why, and why? Using a home problem as an example, why did my daughter not complete her homework?

The 5 whys

1. Why did she not do her homework?
2. Did she fall asleep as soon as she opened her books?
3. Did she stay out too late with her girlfriends?
4. Did I give her permission to go to the mall with her girlfriends?
5. Because my wife and I have a policy that our daughter cannot go out on school nights, did I break that policy?

The first impression is that my daughter was an irresponsible teenager; my final impression after the five "whys" is that I was an irresponsible parent, and she was 13. The corrective action is that I would stick to house policy and not allow my daughter to go out on school nights, and my wife and I would make sure that her homework was completed nightly.

In an organizational concept, the objective evidence that is captured must be the measurement of key indicators. Using the COPQ monthly report in Figure 2.1 for a small company in Los Angeles, you can see the same indicator as used in the example of my daughter's homework problem. The indicator was $54,250 of monthly losses. There was only one shop nonconforming material report (NCMR), but it cost the organization $50,125 this month. The next step would be to apply the five whys to find the root cause and preventive measures. The root cause turned out to be a missed requirement during the contract review and planning review processes. The material had to be forged, with full chemical and mechanical reports. The material was used to meet the chemical and mechanical requirements, but was not forged, rendering the material useless for its intended application.

The next area of interest is the ten discrepancies in the internal audit corrective actions. These need to be looked at closely, and this particular organization reviews two attributes of its quality system each month. The attributes this month were contract review and in-process inspections. Although the total cost for each finding was valued at less than $150 each, it indicates an unstable quality system because of the amount of discrepancies. After the review of the internal audit was completed, there was a direct link to the missed contract requirement of using forging for the customer's contract.

The second part of the process was the in-process inspection system; this function also missed the need for a forging to be used.

It takes a team of workers acting as subcontractors, working independently to review previously completed steps, and possessing the organizational freedom to raise the red flag whenever any problems are suspected, without repercussion or any form of discouragement.

Objective evidence brings to the light what would otherwise be in the dark: "We can be comfortable living with our mistakes, and even good at fixing them. But to admit to our mistakes and take steps to remove the cause of the mistakes is the true answer to success."

Description	Quantity	Waste	Total	In Dollars
Shop NCMR	1	400	401	$ 50,125.00
Customer NCMR	5	6	11	$ 1,375.00
Customer CAR	4	2	6	$ 750.00
Internal Audit Findings	10	1	11	$ 1,375.00
Late Deliveries	3	2	5	$ 625.00
	Total Waste	434	Total Loss	$ 54,250.00

Note:
1. NCMR - Nonconforming Material Report
2. CAR - Corrective Action Report
3. The minimum cost for generating an NCMR or CAR is $125.00.
4. Each # = $125. Ex.: (7 × $125.00 = $875.00) (14 × $125.00 = $1750.00)
5. Cost to operate any piece of equipment is equal to $250 or #2

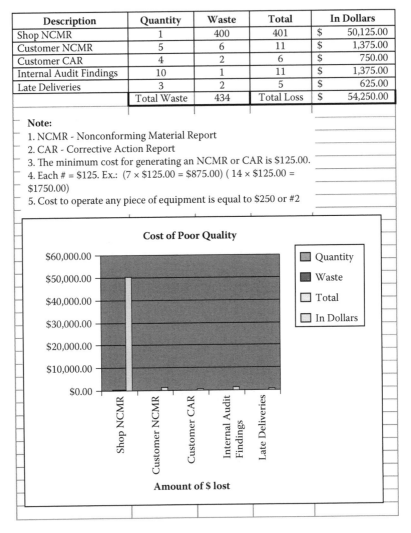

Figure 2.1 Monthly cost-of-poor-quality report.

Chapter three

Contractual requirements

To begin this chapter, I would like to provide you with the legal definition of a contract: (1) offer, (2) acceptance, and (3) consideration.

The following is an example of how a simple contract would be initiated and agreed upon by two parties:

- *Offer.* Bob: I would like you to build me a garage cabinet that will store my tools; the budget for the project is $300.
- *Negotiation.* Ray: I could build that cabinet [6 feet high × 26 inches deep × 36 inches wide] with three shelves that are spaced according to your specifications. The material will be particle board, the cabinet will have two hinged doors with locks, and I will paint it white. The cost for this project will be $375.
- *Counteroffer.* Bob: If you build it for $350 and it is completed within two weeks, I will agree to your terms.
- *Negotiating terms.* Ray: I can do it for $350, and I will deliver it in two weeks, but I will need a $100 deposit.
- *Acceptance.* Bob: That sounds great! I agree on $350, and here is the $100 deposit. I will pay the balance when you deliver my cabinet.

This example illustrates how a simple contract must address the specifications of the project before both parties can accept the contract. If Ray had said that he would build the cabinet without understanding Bob's specifications, the finished cabinet might have been much smaller or much larger than Bob needed. The above contract should have been in writing, not necessarily as a formal contract, but written in the terms as agreed upon by Bob and Ray. The deposit would serve to bind the contract. Consideration is usually required in any amount of legal tender (dollars, gold, stamps, etc.); however, a contract can also involve the exchange of services or products with a certain value.

Contracts come in many forms, including statements of work (SOWs), purchase orders (P.O.'s), long-term contracts (LTCs), and contract deliverables (CDRLs). The name of the contract depends on the local area of work or the community's culture. All contracts must include the offer, acceptance, and consideration.

As defined by standards such as ISO 9001:2008 (International Standards Organization [ISO] 2008), the purpose of contractual requirement reviews is to assign responsibility for products or services ordered by the customer. Contractual reviews can ensure that the customer requirements are fully understood, and that the company responsible for delivering the product or service is able to satisfy the requirements and deliver on time. Contractual review is the most important requirement because the customer possesses the power in the contract it presents to its vendor or subcontractor. The old saying is "Be careful what you ask for," or "Be careful what you don't ask for," because it should all be outlined in the wording and clarity of the contract.

I would like to point out one shortcoming that is common throughout the industry.

In your role as a "new manager of the 21st century," contractual agreements need to be expanded to incorporate contractual requirements that meet not only the customer's specifications, but also those of the customer's organization. If you are the customer or part of the customer's organization, the contract applies to you, and you must have input regarding the contract. The input must be defined in the contract as a configuration item. Remember, the configuration item is a key characteristic that will make or break the relationship that you have with the contractor or contracted organization.

I have spoken about CDRL, a government acronym which stands for contract deliverable. This term refers to items, services, or documents that must be physically delivered or performed before the contract can be paid. The problem is that CDRLs are the contracting department's responsibility to incorporate into the contract between the contractor and contractee, and often do not take into consideration the needs of the other organizations. Although the government has different acronyms, all organizations use the same processes. The person who develops the contract (e.g., the purchase order, attorney contract, request for work, or proposal) and the person writing the contract are typically more concerned with the cost of the item and the terms of agreement than with the act of doing the work or providing the service.

Receiving contracts or requests for quotes and proposals

All inquiries and orders are typically delivered to the sales department for review and processing. One potential problem surfaces when sales personnel are interested in getting the contract and commissions, because they have solicited the work or have been the main contact for the contractor. The sales department may be too quick to agree to the contract before examining the amount of time and materials needed to satisfy the deliverables. This increases the importance of contract review prior to acceptance.

Contract review

The minimum criteria for reviewing customer product requirements prior to contract acceptance are as follows:

- The product requirements must be well defined.
- Contract order requirements differing from those previously expressed must be resolved.
- The organization must have the ability to meet the defined requirements.

- Delivery risks have been evaluated for new technology changes or commitments that may affect delivery times.

Only authorized personnel may respond to customer requests for offers. The sales department can review the requests for completeness and clarity of stated requirements, and note any special instructions for submitting offers. If a discrepancy or ambiguity is noted, the customer is contacted for clarification.

Following this initial review, departments that have a position in the contract such as operations, purchasing, and quality assurance should review the contract. If, during preparation of the offer, it is noted that the customer requirements are incomplete, ambiguous, or impossible to fulfill, the customer must be contacted.

Verification of capacity to fulfill requirements

Before an offer is agreed upon, the availability of materials, man-hours, and resources is determined in order to ensure that the customer's delivery date can be met.

If the company cannot completely fulfill the customer's requirements, the customer is contacted and a modification of the problematic requirement is proposed.

Order changes and amendments

Order changes and amendments received from the customer are routed to the same functions that initially approved the contract or work package. When a change order is received:

- The corresponding original order is retrieved.
- The change order is reviewed for completeness and clarity, and the customer is contacted if more information or clarification is needed.
- The status of the original order processing (accepted, entered, scheduled, in production, packed, or shipped) is determined.
- The feasibility of the requested change and the impact on cost and delivery date are determined, and a change order offer is prepared.
- Upon acceptance of the offer by the customer or special authorization by the sales department, appropriate instructions are issued to implement the change order. Depending on the processing status of the original order, the change order is forwarded to functions concerned (such as accounting, production, purchasing, packaging, and shipping).

When production-scheduling considerations require a change order immediately without waiting for formal acceptance by the customer, the president of the company or his or her delegated representative can consider the risk and authorize the implementation of the change order by writing instructions on a traveler work order that is signed and dated.

Contract review record

A copy of an offer, acknowledged by the concerned departments, constitutes the evidence and record of review. Not only is it important to have evidence of review, but it is also important to have a record of who performed the review and supporting data. The same applies to offers for change orders. The storage location and retention period for the review of records must be specified in the organization's operating procedures. Records are used for history, lessons learned, and accounting.

The review is not a review if you are unable to quantify the review with data estimating the profit margin if completed without mistakes, and the profit margin if completed with reasonable management errors. Most small-to midsized organizations are unable to tell you what the real profit margin is on any given contract.

Reference

International Organization for Standardization (ISO). 2008. *ISO 9000:2008*. Geneva: ISO.

Chapter four

The "lost tribal wisdom" of project management

When did Generation's X and Y "Millennials" lose their mentors and trainers?

I have one client that has a staff of personnel who perform all the shaping and forming of Boeing aircraft frame structures. Everything is done by hand and checked by an employee's eyes. For two years in a row, they were in the 98 to 99 percent rating by Boeing; during my last visit, I reviewed the discrepancies and found one minor administrative problem, and all other problems were mistakes in drawings by Boeing. They have few computers available to control administrative functions. The technician makes the product that the customer pays for, and the customer does not care where the parts come from, as long as they are correct.

This implies that the new manager could be capable of making mistakes of the past. It also means that the contract battles today sound like "You did not tell me how to do it, so why are you telling me it's wrong?" and "I'm not paying because you were not articulate in what you wanted." When you have a good relationship with your subcontractor, meaning they make the changes with a smile, you are paying too much, or you have them over a barrel. Either way you lose. Paying too much for a service is not good business practice. If you have your contractor over a barrel, I can guarantee that they are looking for another customer, and when they find the customer they will drop you like a hot potato, usually when you need their support most.

Missing tribal wisdom means that you have to keep looking for the latest and greatest in technology, software, hardware, and so on. It will get to the point where you will be so engrossed in the "new-newer-newest" syndrome that you will be up to your eyeballs in technology and will lose the knowledge and wisdom base that can only be captured by people. Some look at it as greed; some look at it as advances in technology.

I believe that only about 10 percent of today's industries can fully automate, while 60 percent see it as the carrot to fast profits. Most of the time they believe it is a long-term investment. Unfortunately, considerations for downtime, maintenance and breakdowns, setup time, and a lack of configuration management can eat up profits. Wisdom is the best business defense. I remember as a kid reading a Bible story about King Solomon in the Old Testament. In the Book of I Kings 3:9 (New International version), Solomon asked God, "So give your servant a discerning heart to govern your people and to distinguish between right and wrong. For who is able to govern this great people of yours?" King Solomon was given a discerning heart.

I believe that wisdom is the first requirement for understanding the basic requirements of a design. In the following sections, I will share some basic knowledge required to be able to identify the surface characteristics of metal.

Surface characteristics of metal

While you must control the finished dimensions of a part, you must also consider the degree of smoothness, or surface roughness. Both are very important in the efficiency and life cycle of a machine part. A finished surface may appear to be perfectly flat, but when you examine it with surface finish–measuring instruments, you will find it is formed of irregular waves. On top of the waves are other smaller waves that are called *peaks* and *valleys*, and you'll measure these peaks and valleys to determine the surface roughness measurements of the height and width. The larger waves are measured in order to determine the height and width measurements. Figures 4.1 and 4.2 illustrate the general location of various areas for surface finish measurements and the relation of the symbols to the surface characteristics.

Surface roughness is the measurement of the finely spaced surface irregularities, including height, width, direction, and shape, in order to establish the predominant surface pattern. The cutting or abrading action of the machine tools that have been used to obtain the surface causes these irregularities. The basic roughness symbol that you will find on the drawing is a checkmark. This symbol is supplemented with a horizontal extension line above it when requirements such as waviness width or contract area must be specified in the symbol. A drawing that shows only the basic symbol indicates that the surface finish requirements are detailed in the NOTES block. The roughness height rating is placed at the top of the short leg of the check (Figure 4.3).

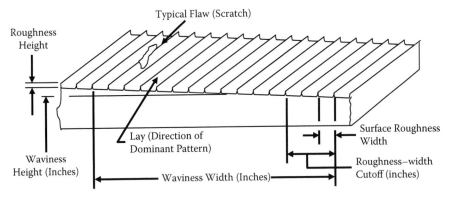

Figure 4.1   The general location of various areas for surface finish measurements, and their relation of the symbols to the surface characteristics.

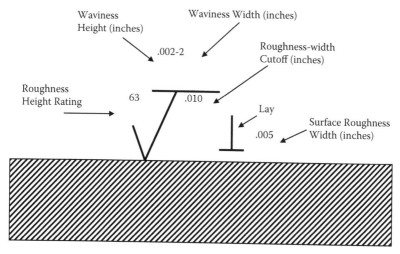

Figure 4.2   The general location of various areas for surface finish measurements, and their relation of the symbols to the surface characteristics.

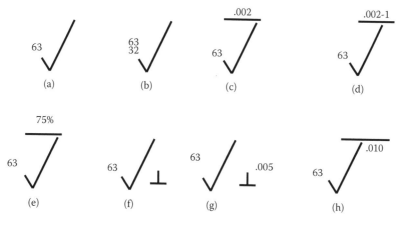

Figure 4.3   Roughness height rating.

If only one number is shown for roughness height, it is the maximum permissible roughness height rating. If two are shown, the top number is the maximum (Figure 4.3, view B). It is important to remember that the smaller the number in the roughness height rating, the smoother the surface.

Waviness height values are shown directly above the extension line at the top of the long leg of the basic check (Figure 4.3, view C). Waviness width values are placed to the right of the waviness height values (Figure 4.3, view D). Where minimum requirements for contact or bearing surfaces must be shown, the percentage is placed at the top of the long leg of the basic check (Figure 4.3, view E). The NOTES block of the drawing will show any further surface finish requirements, such as waviness, width, or height.

Lay is defined as the direction of the predominant surface pattern produced by the tool marks. The symbol indicating lay is placed to the right and slightly above the point of the surface roughness symbol, as shown in view F of Figure 4.3. In Figure 4.4, the roughness width value is shown to the right of the parallel to the lay symbol. The roughness width cutoff is placed immediately below the extension line and to the right of the long leg of the basic checkmark. These symbols for roughness width are shown in views G and H of Figure 4.3.

Figure 4.5 shows a sampling of some roughness height values that can be obtained by the different machine operations.

Reading surface finish quality

A surface finish is seldom flat. As mentioned earlier, close examination with surface finish–measuring instruments shows the surface finish–measuring waves, and on top of the waves are other smaller irregularities known as peaks and valleys. We will now discuss several ways to evaluate surface finish.

- Visual inspection
  - There are occasions when visual comparison with the naked eye will show that one surface is rougher than the other. It is possible only in cases of widely differing surfaces. You also can use visual inspection to detect large cracks in metal.
  - You can make a visual comparison with illuminated magnifiers.
  - Touch comparison: move a fingernail along the surface of the job and make a mental note of the amount of resistance and the depth of irregularities. Then, move your fingernail across a series of master roughness scales that have numbers corresponding to their measurement in microinches (Figure 4.5). The machine finish must compare satisfactorily with the correct master.
- Interference microscope inspection: this requires the use of a microscope with an optical flat plate and a monochromatic light. The microscope allows you to see the height of the surface irregularity in light reflected between the microscope objective and surface of the work.
- Profilometer: this instrument is most commonly used to find the degree of surface roughness. It uses the tracer method and actually measures the differences in the depth of the surface irregularity.
- Surface analyzer: the surface analyzer is a practical shop instrument designed for the accurate measurement of surface finish roughness. Like the profilometer, it measures the irregularities of the surface finish and records them in microinches. This is done by a tracer stylus, which registers the rise and fall of the peaks and valleys on the finished surfaces. These variations are amplified and indicated on the electrical meter, which is calibrated to read in microinches.

Lay Symbols	Designation	Example
=	Lay parallel to the boundary line representing the surface to which the symbol applies	Direction of Tool Marks
⊥	Lay perpendicular to the boundary line representing the surface to which the symbol applies	Direction of Tool Marks
X	Lay angular in both directions to boundary line representing the surface to which symbol applies	X
M	Lay multidirectional	M
C	Lay approximately circular relative to the center of the surface to which the symbol applies	C
R	Lay approximately radial relative to the center of the surface to which the symbol applies	R
P	Lay particulate, non-directional, or protuberant (Not shown in ISO 1302 Standard)	P

Figure 4.4   The seven symbols that indicate the direction of lay.

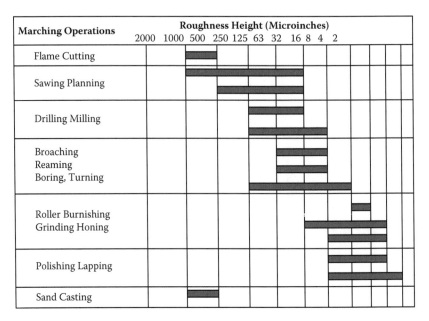

Marching Operations	Roughness Height (Microinches)											
	2000	1000	500	250	125	63	32	16	8	4	2	
Flame Cutting		▓▓										
Sawing Planning		▓▓▓▓▓▓										
Drilling Milling				▓▓▓								
Broaching Reaming Boring, Turning				▓▓▓▓								
Roller Burnishing Grinding Honing							▓					
Polishing Lapping								▓▓				
Sand Casting		▓▓										

Figure 4.5   A sampling of roughness height values that can be obtained by the different machine operations.

Accuracy of equipment and personnel

This section will give you an overview on how to perform gauge R&R (round-robin) methods in your workspace, and why these methods are so important. We calibrate tools, and we know the accuracy of equipment. However, what about the people using the tools? How accurate are your workers? In this section, we will address the following:

1. Limits of accuracy defined on drawings
2. Limits and accuracy of equipment
3. Limits and accuracy of personnel

Limits of accuracy defined on drawings

Understanding of tolerance and allowance

These terms may seem closely related, but each has a very precise meaning and application. I would like to point out the meanings of these terms and the importance of observing the distinction between them.

Tolerance

Tolerances Should Fit Product and Process Need.   Although it is possible by use of sufficient time and care to work as closely to a given dimension as is desired, it is impossible to manufacture to an exact size. Regardless of the accuracy displayed, it is always possible to choose a finer measuring method that can show discrepancies in the dimension. Because working toward higher accuracies increases costs in terms of money, time, and equipment, it is more practical and economical that dimensions should be permitted to vary within the widest limits for which they can still function properly. This variation is permitted by the use of *tolerances* added to dimensions in such a way that they indicate the permissible variation. Theoretically, the designer applies dimensional tolerances as wide as can be safely used.

In most instances, it is impractical and unnecessary to work to the absolute or exact basic dimension. The designer calculates, in addition to the basic dimensions, an allowable variation. The amount of variation, or limit of errors permissible, is indicated on the drawing as plus or minus (+/–) a given amount, such as +/– 0.005 or +/– 1/64. Illustrated in Figure 4.6 is the difference between the allowable minimum and the allowed minimum dimension in tolerance.

When tolerances are not actually specified on a drawing, you can make fairly concrete assumptions concerning the accuracy that is expected by using the following principles. For dimensions that end in fractions of an inch, such as 1/8, 1/16, 1/32, 1/64, the expected accuracy is +/– 1/64. When the dimensions are given in decimal form, the following applies:

> If the dimension is given as 3.000 inches, the accuracy expected is +/– 0.0005 inches; or if the dimension is given as 3.00, the accuracy expected is +/– 0.005 inches. The +/– 0.0005 is called, in shop terms, "plus or minus five ten-thousandths of an inch." The +/– 0.0005 is called "plus or minus five one-thousandths of an inch."

Allowance. Allowance is an intentional difference planned in dimensions of mating parts to provide the desired fit. A *clearance allowance* permits movement between mating parts when they are assembled. For example,

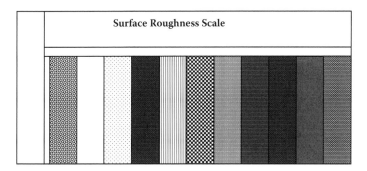

Figure 4.6 The difference between the allowable minimum and the allowed minimum dimension is tolerance.

when a hole with a 0.250-inch diameter is fitted with a shaft that has a 0.245-inch diameter, the clearance allowance is 0.005 inches. An *interference allowance* is the opposite of a clearance allowance. The difference in dimensions in this case provides a tight fit. You would need force to assemble parts that have an interference allowance. If a shaft with a 0.251-inch diameter is fitted into the hole identified in the preceding example, the difference between the dimensions will give an interference allowance of 0.001 inches. As the shaft is larger than the hole, force is necessary to assemble the parts.

What is the relationship between tolerance and allowance? When you manufacture mating parts, you must control the tolerance of each part, so that the parts will have the proper allowance when they are assembled. For example, a hole 0.250 inches in diameter with a tolerance of 0.005 inches (+/− 0.0025) is prescribed for a job. For the shaft to be fitted in the hole, it must have a clearance allowance of 0.001 inches. You must finish the hole within the limits and determine the exact size required for the shaft prior to making the shaft. If you finish the hole to the upper limit of the basic dimension (0.2525 inches), you must machine the shaft 0.2515 or 0.001 inches smaller than the hole. If the dimension of the shaft were aligned with the same tolerance as the hole, there would be no control over the allowance between the parts. As much as a 0.005-inch allowance (either clearance or interference) could result.

To retain the required allowance while permitting some tolerance in the dimensions of the mating parts, the tolerance is limited to one direction on each part. This single-direction (unilateral) tolerance stems from the basic hole system. If a clearance allowance is required between mating parts, the part that fits into the opening may be smaller but not larger than the basic dimension. Thus, the shafts and other parts that fit into a mating opening have a negative tolerance only, while the openings have a positive tolerance only. If an interference allowance between the mating parts is required, the situation is reversed, and the opening can be smaller but not larger than the basic dimension, while the shaft can be larger but not smaller than the basic dimension. Therefore, you can expect to see a tolerance such as +0.005, −0, or +0, −0.005, but with the required value not necessarily 0.005. You can get a better understanding of a clearance allowance, or an interference allowance, if you make a rough sketch of the piece and add dimensions to the sketch where they apply.

Limits and accuracy of equipment

Measurement identification and selection of equipment

The basis for calibration requirements should be compliant with the latest revision of ANSI (American National Standards Institute)/NCSL Z540-1-1994 (MIL-STD [military standard] 45662) and MIL-STD 120, to name a few standards that have been used (see ANSI 1994). When applicable, each gauge or measuring instrument shall be calibrated against a standard whose calibration is certified traceable to the National Institute of Standards and Technology, and has an accuracy ratio of four times greater than the capability of the instrument being calibrated.

Calibration

The department that uses the equipment is responsible for maintenance, calibration, and control of all inspection, measuring, and test equipment, including test devices and tools supplied by the customer. Calibration recall lists are used to schedule and define the need for recalibration. Equipment is calibrated in accordance with the manufacturer's written instructions, unless calibration is simple and obvious. Calibration is conducted at ambient room temperature, typically 68–72°F. When applicable, calibration instructions specify each type of equipment, and the acceptable limits of temperature, pressure, humidity, and other environmental conditions that may affect calibration.

Calibrated measuring and testing are carried out using calibration instruments or standards certified to have a known relationship to a nationally recognized standard. This relationship is identified on the calibration record. Equipment sent out for calibration must be returned with certification that is likewise traceable to a national standard. The department that uses the equipment maintains calibration records and certificates to show objective evidence of its calibration and accuracy.

Calibrated equipment is identified with a serial number and labeled with a sticker indicating the due date for its next calibration. If it is not possible or practical to attach a sticker, the equipment box or container is attached with the sticker. Such equipment is traceable to its equipment calibration record through its serial number. Equipment with a past-due calibration is not used and is reported to the equipment controller.

Reference measuring and test equipment is identified with a serial number and also controlled by the department that uses the equipment. This equipment does not require current calibration or a sticker, only a "for reference only." When the need exists to use reference measuring and test equipment for final acceptance, the equipment will be verified against known calibrated standards using the high-low technique.

The division maintains a list of all active measuring and test equipment. The list identifies every piece of equipment by its description, range, serial number, location, calibration frequency, and last calibration date. The list is updated at least once yearly.

Storage and maintenance

Measuring and test equipment is stored in a secured storage area. The equipment is maintained, stored, and handled in order to preserve its accuracy and fitness for use. Equipment that is out of calibration or is otherwise not fit for use is withdrawn from the inspection and production areas, and delivered to QA for calibration.

Nonconforming equipment

When a piece of measuring or test equipment is found to be out of calibration, or appears to give inaccurate readings, the piece is checked. If it is confirmed that the equipment is indeed out of calibration and the readings are outside of required accuracy, management investigates and assesses the validity of the measurements for which the equipment was previously used. Identification of such equipment and the impact of its use for acceptance of products are dispositioned using a nonconforming report. The investigation may also be concluded with a request for corrective and/or preventive action, and a notification to a customer if the result would affect the form, fit, safety, or function of items inspected using the faulty test equipment.

Equipment exempted from calibration

Inspection and test equipment used in situations where the accuracy of measurements is not important, or where the measurement does not have any relation to the verification of products, is exempted from the calibration requirement. Such equipment is labeled with stickers warning that it is not calibrated. Production and inspection personnel are made aware of the limitations in using uncalibrated equipment.

Limits and accuracy of personnel

Gauge R&R studies are performed to compare the human tolerance criteria of accuracy for a group of personnel. The test is completed in the following steps:

1. Three items are given to each person to measure or test. Only the person performing the test will know the results. The other participants in the test may not see any other person's results.
2. The same three items and measuring tools are given to each person. The disposition is evaluated to create an average and mean, and give the standard deviation of the group

Round-robin testing shall be conducted to assure there is no significant error between operators doing the same test and using the same instruments. If the test results indicate a variation that is not acceptable, additional training in the art of taking measurement is conducted. If necessary, a detailed procedure is established and used until the personnel are within acceptable tolerance limits for the group.

Welding and allied processes

Welding is the permanent union of metallic surfaces through the establishment of atom-to-atom bonds between surfaces. There are a few distinctive characteristics that identify the differences between true welding and brazing or soldering. The main difference is that in welding, the filler material has a similar composition to the base metals that are being welded together, and there is an attempt to join the metals as one. In brazing and soldering, however, the filler metal has a lower melting point than the base metal(s), making it an adhering process. Similarly, in the plastics industry, adhesive joining is sometimes performed as a true welding process with certain plastics, making use of organic adhesives that often contain plastic filters and inorganic solvents that fuse the surfaces of the plastic and adhesive together. With some plastics, introducing a volatile solvent into the joint, which "melts" the plastic-plastic interface, essentially welding the parts together, can form a sound plastic-plastic joint. In this section, we will discuss welding and allied processes in reference to joining metals.

Evidence indicates that when prehistoric humans found native metals in small pieces and were unable to melt them, they created larger pieces by heating and welding, and by hammering or forging the pieces together. Arc welding was first used in 1880, and oxyacetylene in 1895. Arc welding (Figure 4.7) is the process of using an electric current to produce an "arc" to melt two pieces of metal and join them as one. The heat concentration is quick and fast, and when principles of heat control and procedural techniques are used correctly, the metal is joined with little distortion.

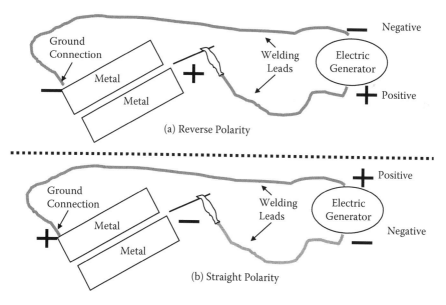

Figure 4.7 Arc welding: the process of using an electric current to produce an "arc" to melt two pieces of metal and join them as one.

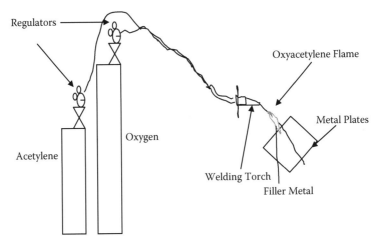

Figure 4.8 Oxyacetylene welding: a nonpressure process in which coalescence (growth together) is produced by heat from an oxyacetylene flame formed by the combustion of oxygen and acetylene.

Oxyacetylene welding

Oxyacetylene welding (Figure 4.8) is a nonpressure process in which coalescence (the act of growing together) is produced by heat from an oxyacetylene flame formed by the combustion of oxygen and acetylene. The two gases are mixed to correct proportions in a torch, and the torch can be adjusted to give various types of flame.

Oxygen is a colorless, tasteless, odorless gas that is slightly heavier than air. Oxygen will not burn by itself, but it will support combustion when combined with other gases. Extreme care must be taken to ensure that compressed oxygen does not become contaminated with hydrogen or hydrocarbon gases or liquids, unless the oxygen is controlled by such means as the mixing chamber of a torch. A highly explosive mixture will be formed in uncontrolled compressed oxygen when it becomes contaminated. Oxygen should *never* come in contact with oil or grease because it can be self-igniting.

There are two major concerns when working with oxygen. First of all, oxygen will support combustion, and it is heavier than air. You have probably seen signs posted at hospitals and other locations prohibiting smoking around oxygen equipment, but why can't we smoke around oxygen? The following is a real-world example to illustrate this concept. In the world of deep underwater diving, there is a process known as *saturation diving*, which occurs when the diver goes to extreme depths under the water. It requires that the diver be subject to a highly compressed atmosphere of oxygen and helium. In this atmosphere if a spark were to occur, all combustible materials would burst into flames. We have lost many saturation divers through both a spark and hydrocarbon (oil or grease) under impact in an oxygen-enriched atmosphere. The hospital environment would most likely have an incident where pure oxygen being administered to a patient would saturate pajamas, or bedcovers, and if subject to spark or flame would most surely ignite the patient through an explosion ending in death.

Oxyacetylene cutting of metal is more common today than oxyacetylene welding. The electric arc welding process is used to weld most metals because it is easier to control the parameters needed to weld different types of metals. In the old days, I was trained to weld aluminum using the oxyacetylene welding technique.

Most welders today will say aluminum cannot be welded by oxyacetylene because, unlike other metals, you cannot see the aluminum color changes that indicate when the metal is ready to be welded. In fact, you cannot see the temperature change by color unless you first use the acetylene gas to blacken the aluminum. When you have the black carbon coating on the weld area, the torch mixture is adjusted to add the oxygen to the right percentage to give you a neutral flame. The neutral flame is about 5900°F. and as you use the torch to heat the weld area the blackened area will turn clear, indicating that the metal is ready to be welded. The remainder of the process is dependent on practice that is required to control the speed of travel and heat control. It is complicated at first, but before the gas metal arc welding (GMAW) and gas tungsten arc welding (GTAW) processes (in the 1960s called the tungsten inert gas [TIG] welding process) were developed, we had no choice when using aluminum.

Oxyacetylene welding has many safety precautions, such as those regarding putting your clothing on fire, gas cylinder safety, and the ones I have seen cause the most damage: backfire and flashback. *Backfire* is a momentary burning back of the flame into the torch tip, and *flashback* occurs when the flame burns in or beyond the torch mixing chamber. Unless the system is thoroughly purged of air and all connections in the system are tight before the torch is ignited, the flame is likely to burn inside the torch instead of outside the tip. A backfire is characterized by a loud snap or pop as the flame goes out. A flashback is usually accompanied by a hissing or squealing sound, and you don't want to hear this sound. At the same time, the flame at the tip becomes smoky and sharp pointed. When a flashback occurs, immediately shut off the torch oxygen valve and then close the acetylene valve. By closing the oxygen valve, the flashback is stopped at once. A backfire is typically less serious.

Welding processes

When it comes to welding processes, there are many differing types, as illustrated in Figure 4.9. All welding processes, with the exception of brazing, use temperatures high enough to melt the base metals. Brazing is the only welding process in which the melting of the base metal is not necessary for

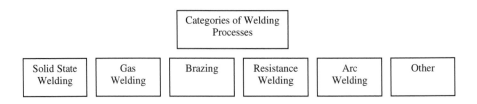

Figure 4.9   The primary categories of welding processes.

coalescence. Brazing is similar to soldering, except that higher temperatures are used for brazing. The term *soldering* is used to describe a joining process using nonferrous filler alloys melting below 800°F. Soldering is not considered a welding process. Brazing is a welding process using nonferrous filler alloys that have a melting point above 800°F but below that of the base metal.

Subcategories of the above number in the hundreds; some subcategories for solid-state welding processes are as follows:

1. Ultrasonic welding
2. Friction welding
3. Forge welding
4. Explosion welding
5. Diffusion welding
6. Cold welding

Subcategories of arc welding are as follows:

1. Stud welding
2. Plasma arc welding
3. Submerged arc welding
4. Gas tungsten arc welding
5. Gas metal arc welding
6. Flux cored arc welding
7. Shielded metal arc welding
8. Carbon arc welding

Subcategories of brazing are as follows:

1. Infrared brazing
2. Torch brazing
3. Furnace brazing
4. Induction brazing
5. Resistance brazing
6. Dip brazing

Subcategories of other welding processes are as follows:

1. Thermit welding
2. Laser beam welding
3. Induction welding
4. Electroslag welding
5. Electron beam welding

The welding processes differ not only in the way coalescence is achieved, but also in their ability to produce a satisfactory joint in a given kind of metal under the conditions in which the weld must be made. Many factors influence the selection of a welding process of a particular application.

Characteristics of metal

The fundamental characteristics of metal include elasticity, deformation, toughness, hardness, weight, appearance, weldability, resistance to corrosion, and resistance to chemicals.

The objectives of this section are as follows:

- To explain the concepts of stress and strain in metals
- To describe the different properties of metals
- To identify the two major classes of metals
- To describe the different types of ferrous and nonferrous metals
- To identify different metals by color markings, surface appearance, and identification tests

What do you know about metals? Chemical elements are considered to be metals if they are lustrous, hard, good conductors of heat and electricity, malleable, ductile, and heavy. Some metals are heavier than others, some are more malleable than others, and some are better conductors of heat and electricity. These properties are known as *metallic properties*, and chemical elements that possess these properties to some degree are called *metals*. Chemical elements that do not possess these properties are called *nonmetals*. Oxygen, hydrogen, chlorine, and iodine are examples of nonmetallic chemical elements.

Chemical elements that behave sometimes like metals and sometimes like nonmetals are often called *metalloids*. Carbon, silicon, and boron are examples of metalloids.

An alloy may be defined as a substance that has metallic properties and is composed of two or more elements. The elements that are used as alloying substances are usually metals or metalloids. By combining metals and metalloids, it is possible to develop alloys that have the particular properties required for a given use.

Figure 4.10 lists some common metals and metalloids and provides the chemical symbol that is used to identify each element.

Element	Symbol
Aluminum	Al
Antimony	Sb
Cadmium	Cd
Carbon	C
Chromium	Cr
Cobalt	Co
Copper	Cu
Iron	Fe
Lead	Pb
Magnesium	Mg
Manganese	Mn
Molybdenum	Mo
Nickel	Ni
Phosphorus	P
Silicon	Si
Sulfur	S
Tin	Sn
Titanium	Ti
Tungsten	W
Vanadium	V
Zinc	Zn

Figure 4.10   Some common metals and metalloids, and the chemical symbol that is used to identify each element.

Iron and steel

Carbon steel is an alloy of iron and controlled amounts of carbon. *Alloy steel* is a combination of carbon steel and controlled amounts of other desirable metal elements.

The percentage of carbon content determines the type of carbon steel. For example, wrought iron has 0.003 percent carbon content. Low-carbon steel contains less than 0.30 percent carbon. Medium-carbon steel varies between 0.30 and 0.55 percent carbon. High-carbon steel contains approximately 0.55 to 0.80 percent carbon, and very-high-carbon steel contains between 0.80 and 1.70 percent carbon. Cast iron contains 1.8 to 4 percent carbon.

Carbon generally combines with the iron to form *cemetite*, a very hard, brittle substance. Cemetite is also known as *iron carbide*. This action means that as the carbon content of the steel increases, the hardness, strength, and brittleness of the steel also tend to increase.

Various heat treatments are used to enable steel to retain its strength at the higher carbon contents, and yet not have the extreme brittleness typically associated with high-carbon steels. Also, certain other substances, such as nickel chromium, manganese, vanadium, and other alloy metals, may be added to steel to improve certain physical properties.

A welder must also have an understanding of the impurities occasionally found in metals and their effect upon the weldability of the metal. Two of the detrimental impurities sometimes found in steels are phosphorus and sulphur. The presence of these impurities in steel comes from the ore, or the manufacturing method. Both of these impurities are detrimental to the welding qualities of steel. Therefore, during the manufacturing process, extreme care must be taken to keep the impurities at a minimum (0.05 percent or less). Sulphur improves the machining qualities of steel, but is detrimental due to its hot forming properties.

During a welding operation, sulphur or phosphorus can form gas in the molten metal. The resulting gas pockets in the welds cause brittleness. Another impurity is dirt or slag (iron oxide). The dirt or slag is imbedded in the metal during rolling. Some of the dirt may come from the by-products of the process of refining the metal. These impurities may also produce blowholes in the weld and reduce the physical properties of the metal in general.

Stress and strain

When external force is applied to any solid material, the material is subject to stress. Many of the properties of metals can best be understood in terms of the manner in which they react to stress. Therefore, before considering the properties of metal and alloys, let's examine the concepts of stress and strain.

Load, which is typically measured in pounds, is the external force applied to a material. When the load is applied, reaction forces to the load occur throughout the material. The reaction forces are stresses. However, why do these forces occur when a load is applied to the material? Newton's third law of motion states that "to every force or action, there is an equal and opposite reaction." Stress, therefore, is the "equal and opposite" reaction to the externally applied load. It is defined as the force per unit area resisting the load. Unit area is important. The unit area may be stated as a square inch, a square foot, or any other predetermined amount of area that is used to figure the amount of stress that material will be subjected to. When the load is applied, it is distributed equally throughout the cross section of the material. For example, suppose two round metal rods with cross-sectional areas of 1 square inch and 2 square inches are each supporting a 2,000-lb. weight. The load or external force is the same on both, but since the cross-sectional areas are different and the load is distributed equally over the cross-sectional areas, the stresses in the two rods are also different.

You can see from this example that stress is equal to the load divided by the cross-sectional area. That is, equal portions of the load are distributed equally over the cross-sectional areas. Stress is usually measured in pounds (for load) per square inch (for area). Conversely, the load can be determined by multiplying the stress by the cross-sectional area.

Strain is the deformation or change in shape caused by the load. Some strain always occurs as a reaction to a load. The amount of strain depends on the magnitude and duration of the stress caused by the load. It also depends on the type and condition of the material. Strain is measured in inches per square inch, or in percentages. Thus, when a load is applied to a bar in tension, the bar will elongate (be strained) some fraction of an inch for each inch of bar (the strain will be the same in each inch of bar). If strain is being measured in percentages, the bar will be elongated a certain percentage; that is, the total length of the bar will be increased a certain amount, which will be a percentage of the original length.

Stress occurs because molecular forces within the material resist the change of shape that an applied load tends to produce. In other words, stress results from the resistance of the molecules being shifted around, pulled apart, or squeezed together. Because stress involves molecular forces, a piece of metal that is subjected to a load develops an enormous number of stresses, rather than just one stress. If you had more than a very few molecules, you would have to draw thousands or perhaps millions of arrows to indicate all the molecular forces involved. We often speak of stress as though it were one internal force and an action in one direction, that is, the direction opposite to the direction of the applied load. In other words, we consider the *total effect* of all the molecular stresses rather than trying to consider each set of molecular stresses separately.

The manner in which the load is applied determines the type of stress that will develop. Applied forces are usually considered as being of three basic kinds: tension (or tensile) forces, compression forces, and shearing forces. The basic stresses, therefore, are tension (or tensile) stresses, compression stresses, and shearing stresses. Complex stresses such as bending stresses and torsional stresses are combinations of two or more of the basic stresses.

Tension stress

Tension stress develops when a material is subject to a pulling action. For example, if a cable were to be fastened to an overhead clamp and a weight attached to the free end, tension stresses would develop within the cable in order to resist the tension forces that could pull the cable apart. Figure 4.11 illustrates tension forces and the resulting "equal and opposite" tension stresses.

Compression stresses develop within a material to oppose the forces that tend to compress or crush the material. A column that supports an overhead weight is said to be in compression, and the internal stresses that develop within the column are compression stresses. Figure 4.11 shows compression forces and compression stresses.

Shearing stress

Shearing stresses develop within a material when opposite external forces are applied along parallel lines in such a way that they can cut the material. Shearing forces tend to separate material by sliding part of the material in the opposite direction. The action of a pair of scissors is an example of shear forces and shear stresses. The scissors apply shear forces, and the material being cut resists the shear forces by its internal shear stresses. Forces tending to produce shear in a rivet are illustrated in Figure 4.12. Shear stresses are not shown, since they are considerably more complex than tension stresses and compression stresses.

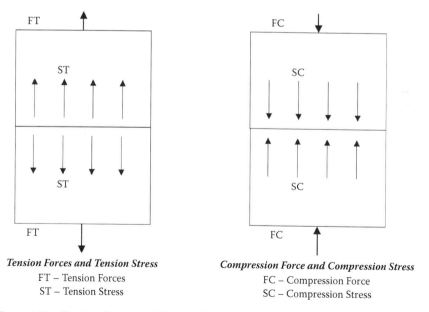

Tension Forces and Tension Stress
FT – Tension Forces
ST – Tension Stress

Compression Force and Compression Stress
FC – Compression Force
SC – Compression Stress

Figure 4.11 Tension forces and the resulting "equal and opposite" tension stresses.

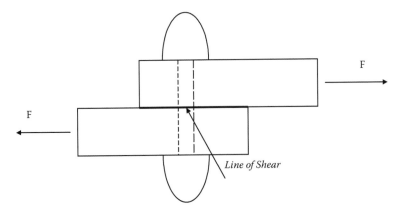

Figure 4.12 Forces tending to produce shear in a rivet.

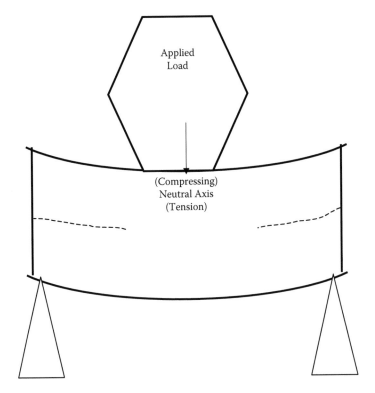

Figure 4.13 When a load is applied to a beam, the upper surface is in compression and the lower surface is in tension.

Bending stress

Bending stresses develop when a material is subjected to external forces that tend to bend it. When a load is applied to a beam, for example, as shown in Figure 4.13, the upper surface is in compression and the lower surface is in tension. The *neutral axis*, indicated by the broken line, is neither in compression nor in tension.

Elasticity

All metals are elastic to some extent. It may surprise you to learn that a piece of steel is more elastic than a rubber band. The rubber band stretches more than the steel since it is more easily strained, but the steel returns almost to its origninal shape and size, becoming truly elastic. Glass is also more elastic than rubber.

The greatest stress that a material is capable of withstanding without becoming permanently deformed is known as the *elastic limit*. Below the elastic limit, the amount of strain is directly proportional to the amount of stress, and therefore to the amount of externally applied force. Above the elastic limit, the amount of deformation that results to form an increase in load is disproportional to the increased load. Therefore, the ratio of stress to strain is a constant for each material. This constant is called the *modulus of elasticity*, and is obtained by dividing the stress by the strain, which is the elongation caused by the stress. For example, imagine that a certain material is loaded to the point that the internal stress developed by the tension is 30,000 pounds per square inch (psi), and that with this stress the material elongates or is strained 0.0015 inches per square inch. The modulus of elasticity (E) of this material is illustrated in Equation 4.1:

The modulus of elasticity (E):

Stress (psi)

E = Elongation (inches per inch)

30,000 psi = 0.0015 inches per inch

Answer = 20,000,000 psi   (4.1)

The modulus of elasticity is frequently used to determine the amount of elongation that will occur when a given stress is developed in the material. For this purpose, you divide the stress by the modulus of elasticity to obtain the elongation (inches per inch) that will occur.

Closely related to the elastic limit of a material is the *yield point*. The yield point is the stress at which deformation of the material first increases markedly without any increase in the applied load. The yield point is always somewhat above the elastic limit. When the stresses developed in a material are greater than the yield point (or, as it is sometimes called, the *yield strength*), the material is permanently deformed.

The lost art of gray cast iron repair

The life expectancy of an engine today can be 200,000 miles or better. The question of rebuilding an engine or purchasing a new or used (or "pre-owned," as dealers prefer) car depends on your budget. In some cases, it makes sense to rebuild, particularly if you are in love with the car you own.

I was recently engaged to help a company that was established in the 1950s to improve the process of engine cylinder head repair. I was amazed to see one of the most efficient processes I have seen in years. Two things can be attributed to the success of this company: (1) the original owners of the company still worked on the shop floor, and (2) this was one case where technology could not replace good old-fashioned, commonsense, case-by-case decisions.

Gray cast iron repair is one of my favorite processes. However, I have not seen it in action since I worked shipboard in the Navy Repair Department in the early 1960s. I have captured this tribal knowledge in one of my training methodologies, as outlined below.

First, we must review the facts that are the foundation of "gray cast iron." This is the first step of any process: understanding the characteristics of the material or product.

Facts about gray cast iron:

- The microstructure of gray cast iron can vary, even within a single casting, and compositions used may vary in tensile strength from 20,000 to 60,000 psi (160 to 250 Brinell hardness number, or BHN).
- Gray cast iron is not suitable for arc-welding repairs, because more cracking will take place.
- The braze-welding process (bronze base filler metal) can be used, but it is not suitable for any part that will be subject to high temperatures, such as auto and truck engines. Braze welding takes place over 800°F with no melting of the base metal (engine head or block) during the braze weld process.
- Areas subject to cylinder pressure cannot be repaired. The valve seat area can be repaired, but not the engine cylinder or water jacket. Both the compression from the cylinder and/or the water coolant could leak.
- The most reliable method of crack repairs in gray cast iron is the plug repair method, another lost art.
- Cast iron is not a single material; in general, it is *not difficult to tap*. Cutting speeds vary from 90 feet per minute for the softer grades to 30 feet per minute for the harder grades. The chip is discontinuous, and *straight fluted taps* should be used for all applications. Oxide-coated taps are helpful, and this cast iron can usually be tapped dry, although water-soluble oils and chemical emulsions can be used.

Cast iron cylinder heads

Cast iron engine parts have been made for years; they were used in the first automobiles that were produced, and are still used today. The cast iron is aged to eliminate expansion and contraction from heat to ambient temperature. How is aged cast iron made?

For *aged cast iron*, the first process is to pour an ingot of cast iron. Then the cast iron is machined to the rough size; the part will be machined to the final dimensions after the aging process. The aging process was discovered in the beginning use of cast iron. People found that as the cast iron expanded and contracted over time, the expansion and contraction would be less and less. Early cast iron engine parts were put in the back lot unprotected to rust and age for about three to five years. What people found was that the cast iron became stable and they could use it as an engine part without worrying about expansion under high temperatures.

Today we artificially age or season cast iron in days by changing the temperature and humidity continually; the result is a cast iron part that can be stabilized in days rather than years.

Cast iron is rigid, and is used mostly for its compression strength and its resistance to expansion and contraction.

Fatigue cracking: cylinder heads

Rebuilding cylinder heads from used cylinder heads creates a few problems. Fatigue crack factors are commonly caused by age, high temperature, expansion and contraction, cyclic action, and the stress riser or weakest point.

Gray Cast Iron + Engine Combustion Cycles + Time + Weakest Area =
Stress Cracking

In-service inspections are performed by visual (VT) and magnetic particle (MT) inspection to find in-service cracks. The cracks are in predictable areas such as the valve hole area. The valve hole is a prime area for cracks because it has thick and thin sections, with pounding from the valves at high temperatures. Cracks propagate (start) from the weakest point (thin and/or abrupt changes in thickness) into the castings.

Cracks in cylinder heads
- Figure 4.14 illustrates fatigue (areas that are worked to a weak point) cracks.
- Magnetic particle testing inspection is used to find cracks in material that is magnetic.
  1. Visible dry powder (see Figure 4.22 as an example of visible dry powder method)
  2. AC yoke or permanent magnet
  3. Longitudinal magnetization

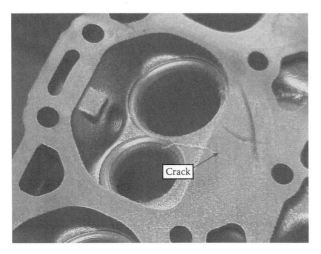

Figure 4.14 Fatigue cracks (areas that are worked to a weak point) indication found by magnetic particle visible dry powder inspection.

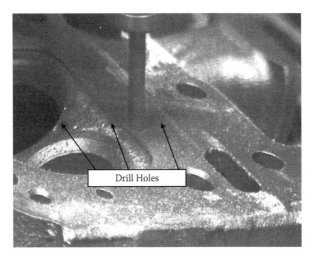

Figure 4.15 Drill holes for tapping (putting internal pipe threads) starting at the crack start and stop points.

Seven Steps to plug-repair a cast iron crack

1. Locate the crack, and center-punch the start, middle, and end of the crack so there is a defined drill start point. This will also guide the drill, preventing the tip of the drill from walking off center and missing the crack start or stop.

2. Drill holes for tapping (putting internal pipe threads; see Figure 4.15), starting at the crack start and stop points. This eliminates the ability of the crack to proergate (continue) after the repair.
3. Tap holes with pipe threads (Figure 4.16).
4. Insert plugs, and torque to desired feet per pound (ft./lbs.; Figure 4.17). The plugs are cast iron to match the base metal composition.

Figure 4.16 Tap holes with pipe threads.

Figure 4.17 Insert plugs and torque to desired feet per pound (ft./lbs.).

Figure 4.18 Using a burring tool, score the plug; use a small hammer and break off excess.

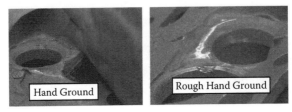

Figure 4.19 Rough-grind and shape the surface so the parts can go through the normal repair procedure.

5. Using a burring tool, score the plug; use a small hammer and break off excess (Figure 4.18). The scoring is used to ensure that the break is above the base metal surface.
6. Continue the insertion of other stitching plugs as necessary, overlapping the others to remove the entire crack area (Figure 4.19).
7. Rough-grind and shape the surface so the parts can go through the normal repair procedure (Figure 4.19).
8. Enter the part into the repair process. Processing is done in groups of the same products, usually 100 to 200 heads per run. This eliminates wasted time to retool for each product run (Figure 4.20).

Weld joint design

Weld symbols and joint symbols are important in finalizing project preparations. This section will outline the American Welding Society (AWS) MIL-STD weld symbols.

The disregard of symbols and end preparations in accordance with the applicable standard, such as ANSI AWS or MIL-STD, can be devastating to configuration management. Many commercial and government drawings only have the welding symbol. Drawings need details to clearly indicate the welded joint and the preparation of material required to achieve the strength and percentage of penetrations.

A common weld symbol is illustrated in Figure 4.21 with and without any references to the military (MIL-STD 1688 and MIL-STD 22, T1V.2) and commercial (ANSI AWS American D.1.1, BTC-P4) standards. As you can see, without the weld joint design symbol, the joint symbol is ambiguous. The engineer can argue that whatever he or she decides to do is right, but he or she

Figures 4.20 Processing is done in groups of the same products, usually 100 to 200 heads per run.

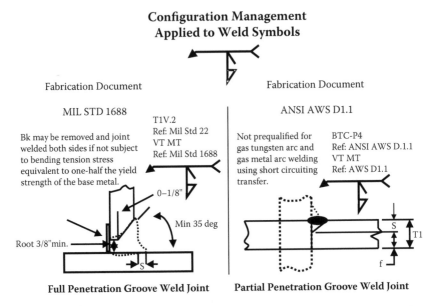

**Configuration Management
Applied to Weld Symbols**

Fabrication Document   Fabrication Document

MIL STD 1688   ANSI AWS D1.1

Bk may be removed and joint
welded both sides if not subject
to bending tension stress
equivalent to one-half the yield
strength of the base metal.

Not prequalified for   BTC-P4
gas tungsten arc and   Ref: ANSI AWS D.1.1
gas metal arc welding   VT MT
using short circuiting   Ref: AWS D1.1
transfer.

T1V.2
Ref: Mil Std 22
VT MT
Ref: Mil Std 1688

0–1/8"

Min 35 deg

Root 3/8"min.

S

S

T1

f

Full Penetration Groove Weld Joint   **Partial Penetration Groove Weld Joint**

Figure 4.21   Common weld symbols with and without any references to the military
(MIL-STD 1688 and MIL-STD 22, T1V.2) and commercial (American National Standards Institute [ANSI] American Welding Society [AWS] D.1.1, BTC-P4) standards.

could be compromising the configuration management and, most important,
the form, fit, function, or safety. In legal terms, the designer has assumed the
liability of the manufactured item without accepting the responsibility and
authority to control the design control configuration.

Here are a few examples of what could be affected:

- The strength of the joint in relationship to the percentage of penetration.
  A full penetration T1V.2 with 100 percent joint efficiency or a partial penetration groove weld BTC-P4 with less than 20 percent joint efficiency.
- A weld with the weld strength T1V.2 equal to the thinnest member, or a
  weld BTC-P4 with no real strength relationship to the base materials.

Heat treatment of metals

The words *heat treatment* mean just about what they say: treating metals with
heat. Heat can be put into metals in a number of ways—by placing work in a
specially designed furnace, by heating with a torch, or by heating with electricity. By far, the most common method is to use a furnace for heating. Heat
can be used to produce a number of different results in metal parts, and this
has led to the design of many different kinds of heat-treating furnaces. Some
may be heated with oil, some with gas, and others with electricity. Some furnaces are designed to heat different types of metals and parts in a variety of

different ways; and others are limited to heating one type of metal or part using a single method.

Heat can be used to do a number of different things to different metals. It can be used to soften a part that is too hard, or it can be used to harden a piece that is not hard enough. Heat can also increase the strength and toughness of a part, it can put a very hard skin on parts that would otherwise be soft, and it can be used to make good magnets out of ordinary material. In short, heat is used to put metals into a condition in which they are useful.

Heat-treated parts are used in many different ways, including saws and axes; all kinds of cutting tools; ball and roller bearings; camshafts and crankshafts; springs, gears, magnets, and fasteners; and seatbelt parts, axles, gun parts, missile cases, and many others. Without heat-treated parts, it would be impossible to build an automobile, a railroad car, an airplane, a watch, a gun, a spacecraft, an earth mover, a bicycle, an outboard engine, or a typewriter. Almost everything we do in our daily living is affected in one way or another by the heat-treating process.

Types of heat-treating processes

Heat-treating processes may be divided into two general types:

1. One that permits or helps a change to take place.
2. One that aims to prevent a change from taking place. The "change," which is either to be helped or prevented is the rearrangement of the tiny particles (atoms) of which the metals are made, this being accomplished without affecting the size and shape of the part being treated.

All heat-treating processes require three steps:

1. Heating to a certain temperature
2. Holding that temperature for a defined period of time
3. Cooling to room temperature

Specifications describe the results that are expected for a particular part. The metallurgist will set up a process that produces these results. Some processes will require very fast heating, some will require very slow heating, and for many the rate of heating is not important. The temperature to which work is heated is most important, and has a direct bearing on the results expected. Temperatures can run as high as 2375°F for some materials, and some special jobs will require that work be "heat treated" 120°F below zero. Time at temperature will vary from a few seconds to as long as 60 hours, depending entirely on the desired results.

The rate at which metals are cooled from high temperature has a very important effect. The fastest ordinary cooling rate comes from a cold-water spray under high pressure. The slowest ordinary rate comes from leaving the work in the furnace and letting the furnace cool at a controlled rate. Fast cooling, called *quenching*, is done in some kind of liquid. Many different types of liquids are used for quenching, such as brine (ordinary salt in water), oils of various types, certain liquid salts, and certain organic materials mixed with water. These liquids provide a variety of cooling rates for quenching, and make it possible to quench every material at the rate that is correct for that material.

Every heat-treated part is expected to be hard enough or strong enough to do a special job. The heat treatment engineer or metallurgist tells the heat treater how to process each part, and every step is important. It is only by carefully following directions that the heat treater can be confident that he or she is making a good part. Poorly treated parts can cause serious accidents.

Materials that are heat treated

Steel is the material that is most commonly heat treated, but there are also some alloys of aluminum, copper, and titanium that can be heat treated. Steels are alloys of iron combined with other materials, the most important of which are carbon steels that have very little carbon and are heat treated to produce a very hard skin on the surface. Steels that have higher carbon are heat treated to make them either very strong or very hard.

There are many different kinds of steels that differ by composition. *Plain carbon steels* are primarily iron plus carbon. *Alloy steels* are iron plus carbon plus one or more metals that have been mixed in while the liquid steel was being made. Some of the common alloy steels may contain manganese (Mn), chromium (Cr), nickel (Ni), or molybdenum (Mo, or moly).

The amount of carbon in any steel determines how tough, strong, or hard it can be made by heat treating. In both plain carbon and alloy steels, the amount of carbon and the kind of alloy in the steel are important in determining how the heat treating must be done. The temperature, time at temperature, and method of cooling all depend on the type of steel.

There are so many different types of steels that a numbering system has been set up to make it simple to describe a type of steel without using a lot of words. The Society of Automotive Engineers (SAE) and the American Iron and Steel Institute (AISI) use four numbers to describe many of the most widely used steels. The first two numbers indicate the type of steel, and the last two indicate how much carbon is present. For example, in SAE (or AISI) 1040, the *10* indicates that this is a plain carbon steel, and the *40* indicates that it has 40 "points" of carbon. A point is a simple way to describe a quantity. SAE 1080 has 80 points of carbon, which is twice as much carbon as in SAE 1040, but both are considered plain carbon steels.

Because the first two numbers of an SAE or AISI number indicate the type of steel, each steel will have its own set of numbers. Table 4.1 illustrates some of the most commonly used steels during a heat-treating operation.

Using Table 4.1 to identify various steels, the SAE 4140 steel would be a chromium-moly steel with 40 points of carbon, the SAE 52100 is a chrome

Table 4.1 Some of the Most Commonly Used Steels during a Heat-Treating Operation

10: plain carbon	46: nickel-moly
11: free cutting	52: chromium
31: nickel-chromium	61: chrome-vanadium
41: chromium-moly	86: nickel-chromium-moly

steel with 100 points of carbon, and the SAE 1144 is "free-cutting" steel with 44 points of carbon. Free-cutting steels are mostly iron plus carbon plus manganese plus sulfur, and are named *free cutting* because they are easy to cut and they form chips instead of long shavings when a hole is drilled.

Tool steels are steels made for a special purpose, and irons are primarily used to do some kind of work on other materials; and while they are not a part of the SAE-AISI numbering system, they are divided into a number of types, with a *letter* used to describe each type, as in Table 4.2.

These types of tool steels are described by one of the following:

1. the kind of work they are expected to do
2. what they are made of
3. the kind of heat treatment they need

Thus, type-S steels are used for hammering other materials, such as in breaking pavement or concrete. *Hot work steels* are used for hammering or squeezing metal parts that are hot. *High-speed steels* are used for cutting metals at a high cutting speed, as in a drill press or lathe. *Water-hardening steels* must be quenched from high temperatures into water in order to become

Table 4.2 Tool Steels Divided into a Number of Types (a Letter Is Used to Describe Each Type)

Type	Kind	Type	Kind
W	Water hardening	H41–H43	Moly hot work
S	Shock resisting	T	Tungsten high speed
O	Oil hardening	M	Moly high speed
A	Air hardening	L	Low alloy, special purpose
D	High carbon, high chrome	F	Carbon-tungsten
H11–H16	Chromium hot work	P	Low carbon mold
H20–H26	Tungsten hot work	(Several miscellaneous)	

hard. *Oil-hardening steels* will become hard by cooling from high temperatures in oil, which is slower cooling than water.

Tool steels are further subdivided into individual steels, which are given a number in addition to the letter. Thus, W1, W2, and W3, while water-hardening steels, are different from one another. A3 and A4 are different steels, but will both become hardened by cooling from high temperatures in air.

Three numbers describe *stainless steels*. The 300 series steels are made of iron plus nickel plus chromium and cannot be hardened by heat treating. The 400 series steels are made of iron plus chromium, and most of them, including 416 and 440, can be hardened by heat treating.

Numbers also describe alloys of aluminum, but the numbering system is not organized. Castings, which are poured to shape in a mold, are described by three numbers—319, 355, and 356. Each of these is made of aluminum plus a variety of other metals. Other alloys that have been hammered, squeezed, or rolled to shape (wrought) have four numbers. For example, 2024, 6061, and 7275 are different wrought alloys that are commonly heat treated.

Alloys of titanium are described by a combination of letters and numbers that indicate what is in the alloy. For example, the alloy Ti-2Fe-2Cr-2Mo is a titanium alloy containing approximately 2 percent iron (Fe), 2 percent chromium (Cr), and 2 percent moly (Mo).

Furnaces used in heat treating

Most heat treatment is done in furnaces. Because there are so many different types of heat treating to be done, and because there are so very many different sizes and shapes of parts to be heat treated, there are many different kinds of heat-treating furnaces. Natural gas is available at a reasonable price; therefore, most heat-treating furnaces are heated with gas. Where natural gas is not available, special-purpose electric power is used to provide heat. Oil is also used for heating some furnaces, as is propane. Propane is a manufactured gas somewhat similar to natural gas, but richer in heating value. Propane is purchased in liquid form and must be stored in large outdoor tanks similar to the smaller tanks seen in farmyards and other rural locations.

Fuel-fired furnaces

All furnaces heated with oil, gas, or propane require that air be mixed with the fuel before it can be burned. No fuel of this type can burn without air. During this process, air and gas are brought together in a mixer, and then placed on a burner where the mixture is burned to produce heat. This is the same action that takes place in an automobile engine, where air and gasoline are mixed in the carburetor and later burn in the cylinder. The only exception is that in this case of fuel-fired furnaces, the burning is very fast and is called an *explosion*. Air and gas must be mixed in the right quantities to get a good flame: too much air in the mixture produces a "lean" flame, whereas too little air produces a "rich" flame (like choking an engine), and neither scenario is good. The mixing unit always must have an adjustment on it to provide the right mixture. Adjusting a mixer to give exactly the right flame is a job for an expert.

When the mixture of fuel and air is burned, the flame makes a new mixture of gases that contain the heat. The action of burning is called *combustion,* and the hot gases coming from the flame are called *products of combustion.* It is the products of combustion that actually carry the heat to the work that is in the furnace. The amount of heat that is produced is regulated by a valve, usually in the airline, which controls the amount of gas-air mixture going to the burners.

The easiest way to heat parts in a furnace is to let the products of combustion pass over or through a pile of the parts to be heated. Many furnaces, called *open-fired furnaces,* are built to do just that. After passing over the work and giving up some heat to the work, the products of combustion pass out of the furnace through a hole called a *vent* or *flue.*

There is a downside to open-fired furnaces—when steel parts are hot, the surface is sensitive. The products of combustion will attack the surface and will do one or both of two things:

1. Form scale.
2. Remove carbon and leave a carbonless skin.

Scale roughens the surface of steel and may actually remove enough of the material to change its size. Removing carbon is called *decarburization (decarb* for short), and the steel is said to be *decarbed.* Because the decarb has little or no carbon, its surface cannot be made hard. This is fine for some types of work, and much heat treating is done in open-fired furnaces. However, for parts that have been machined before heat treating, or for parts that must have a smooth, clean surface after being heat treated, an open-fired furnace is not appropriate. This kind of work must be done in the kind of furnace that will not let the products of combustion touch the work that is in the furnace.

There are two ways to protect work from the products of combustion. First of all, furnaces can be built that have a gastight metal shell in which the work is placed. This shell is called a *retort* or *muffle,* and is open at one or both ends to receive the work. The products of combustion pass over and around the muffle for heating purposes. The muffle becomes hot and, in turn, passes on its heat to the work inside. Thus, the products of combustion are delivered inside the furnace, heat the muffle, and pass out through a vent. They never do come into contact with the work.

Work can also be protected from damage by the products of combustion by heating with radiant tubes. Radiant tubes are made out of metal or porcelain, are 2 × 5 inches in diameter, and pass entirely through the furnace. Each tube has a burner on one end and is vented at the other end. The products of combustion heat the walls of the tube, which become a brilliant red (radiant) and pass on the heat to the work, frequently with the help of a fan inside the furnace. Some radiant tubes are straight, some are U-shaped, and some are W-shaped. In some furnaces the tubes are mounted horizontally, and in others they are mounted vertically.

Furnaces that contain liquid salts are used for heating some kinds of work. Gas-fired salt furnaces are typically called *pot furnaces* because the salt is held in a metal pot. Heating of the pot and the salt in it is done by burners that direct the products of combustion at or around the pot, and then pass out through the vent.

Furnaces heated by electricity

In electric furnaces, heat is produced by passing electricity through bars, coils, or ribbons (different forms of heating "elements") that are located inside the furnace. The electric current passing through the heating elements brings them to a high temperature, just as it does in an ordinary kitchen toaster. Heat from the elements does not have a harmful effect on the surface of the steel, so retorts or muffles are not needed. Electrically heated salt furnaces have one or more pairs of electrodes immersed in the salt. The electricity passes through the salt from one electrode to the other, and heats the salt. Switches in the power line leading to the elements control the amount of heat in a furnace.

Furnace control

All furnaces must keep close control of the temperatures at which they operate. The temperature in any furnace is controlled by a pyrometer (the instrument with the temperature scale) that is connected by wires to a thermocouple. The thermocouple (*couple* for short) is located in the furnace near the work that is being heat treated. The pyrometer is set at the temperature desired in the furnace, and when the furnace temperature drops below the desired temperature, the couple sends a signal to the pyrometer, which then calls for more heat. If the furnace temperature is too high, the couple sends its signal to the pyrometer and the heat is turned off. A *pyrometer* is a delicate instrument and must be checked regularly for accuracy.

Types of furnaces

We will leave you to the joy of getting acquainted with the various types of furnaces in your own shop. However, here is a thumbnail description of some of the most commonly used heat-treating furnaces.

Furnaces that operate above 1200°F

Most of the following types of furnaces may be heated by open-fired burners, radiant tubes, or electricity. All furnaces have a steel outer shell and are lined with a type of brick that resists heat (refractory brick) and helps to keep the heat inside the furnace.

1. *Simple box furnace.* This is what its name indicates—a box-shaped furnace with a hanging door on one or both ends. You could pick up the
smallest one, which is the size of a shoebox, in one hand. Larger ones
may be as much as 10 or 15 feet on one side. Work is placed into and out
of these furnaces by hand using tongs or with the help of some kind of
power loader.

2. *Car-bottom furnace.* This is a box furnace, typically large in size, with a floor
that is actually a flat-topped car running on rails. Work is loaded onto the
car while it is outside the furnace. The car is then rolled into the furnace.
Car-bottom furnaces are among the largest furnaces that are made; some
are large enough to house a railroad boxcar with room to spare.

3. *Carbonitrider, or controlled atmosphere integral quench furnace.* This is a
box furnace that is gastight, and has a quench tank built as a part
of the furnace. Work is placed in baskets, which are pushed into the
furnace. After the work has been at a certain temperature for the
desired amount of time, the basket is pushed out onto an elevator
that drops it into the quench tank. This tank typically contains oil for
moderately fast cooling. The quench tank is totally enclosed, and the
basket is advanced to the quench without ever being exposed to the
outside air. Many different kinds of heat treating can be done in this
type of furnace, and there is likely to be at least one in every commercial heat-treating shop. This is one type of furnace that is never heated
with open-fired burners.

4. *Vacuum furnace.* A vacuum furnace is a furnace built entirely inside
a vacuum chamber. All vacuum furnaces are electrically heated, and
some have a quench tank included in the vacuum chamber. Work can
be heated and cooled in that vacuum, and comes out of the furnace
bright and shiny. Vacuum furnaces are quite expensive to buy and to
run. They are used to heat treat a special kind of material.

5. *Pusher furnace.* A pusher furnace is a long box furnace that has room
for one or two rows of trays that hold the work. The trays are pushed
through the furnace by an air cylinder or other power unit. Some
pusher furnaces deliver the hot work outside the furnace, but most
have a quench system at the exit end so that the work, tray and all, is
quenched before coming out of the furnace.

6. *Belt furnace.* A belt furnace is also a long box furnace, and has an end-
less belt running the length of it. The belts are made of alloy wire mesh,
or of cast alloy parts. The belt runs all the time, and the parts are carried through the furnace on the belt. Some belt furnaces have a quench
tank on the exit end, and parts drop off the belt into the quench; some
have a fairly long cooling chamber for keeping parts clean while they
are cooling; and in other belt furnaces, the parts drop off the belt and
come out of the furnace while they are hot.

7. *Shaker furnace.* A shaker furnace is a long box furnace that has a metal
pan running the whole length of the furnace. The pan is given a length-

wise shaking or bumping action that makes the parts slide the length of the furnace and drop off the far end into a quench tank.

8. *Rotary furnace.* A rotary furnace is a barrel-shaped furnace built around a retort. The retort is a tube that is closed on one end and fitted with a plug on the other. It is mounted on rollers so that it can be kept rotating while being heated. The entire furnace body can be tilted up so that work can be dropped into the open end of the retort (plug removed). At the end of the heating time, the furnace is tipped down and the parts drop out of the retort into a quench tank. A continuous rotary furnace has a screw or spiral built inside the retort. Parts are fed into one end of the retort; the spiral pushes them along as the retort turns, and they drop off at the far end into a quench tank.

9. *Pit furnace.* Pit furnaces get their name from the fact that they are typically located in pits in the floor. They may be round or box shaped, but all of them have their door on the top. Work is loaded onto fixtures, and the loaded fixtures are placed in the furnace with an overhead crane. Pit furnaces do not have a built-in quench, and the load must be lifted out of the furnace with the crane, and then quenched in a separate tank.

Furnaces that operate below 1200°F

Furnaces that operate below 1200°F are made in many series. They may be heated with fuel or electricity, and are typically "open fired." Radiant tubes or retorts are rarely used because products of combustion are not particularly harmful to steel parts at low temperatures.

Low-temperature furnaces include the following:

- *Simple box furnaces* with a front-opening door.
- *Continuous belt furnaces.*
- *Pit-type furnaces* with a door at the top. Note that smaller pit-type furnaces may not need to be located in a pit below floor level.

Most low-temperature furnaces have separate chambers in which the burners or heating elements are located. All have fans to blow the heat from the heating chamber into the work chamber, through and around the work being heated.

Low-temperature furnaces are commonly called *draw furnaces*. This comes from an old expression, "Draw the temper," meaning to heat a brittle quenched part to a low temperature in order to reduce its brittleness.

The origins of discontinuities in metals

Basic training in how metals are made and processed is the foundation of "lost tribal wisdom." This wisdom can identify a metal by its family, its amount of carbon content, the presence of elements such as nickel, and whether the metal was heat treated and how hard it is—all without costly spectrum analysis.

First let's look at the word *discontinuity*. The word simply means a break or interruption in the normal physical structure of an article. A discontinuity in metal may be a hole, a crack, a flaw, or anything else that breaks the continuity of the metal. Discontinuities can be found on the surface of the metal, or within the metal itself.

It is necessary to understand why discontinuities are found in materials. To do this, we are now going to discuss the refining processes that transform various mined ores into usable materials. We are also going to discuss the various metal-forming processes in order to understand why specific discontinuities take the shape that they do.

The smelting and forming processes are the guiding factors in determining the types of discontinuities and where they may be expected to appear in an article as a result of a specific forming process.

Because the cause of discontinuities in all metals is similar, we will discuss only the processing of steel. Some discontinuities found in ferrous metals begin at the steel mill when the iron ore is melted in blast furnaces. *Ferrous* is defined as "of, or pertaining to, iron." Thus, steel is a combination of materials, most of which are derived from iron.

The steel-making process begins with iron ore, coke, and limestone, which are all fed into the top of a blast furnace. As the coke burns, an intense heat is created which removes the oxygen from the iron ore and allows the molten metal to trickle to the bottom of the furnace.

The limestone gathers the impurities in the iron ore, and the impurities become liquid. Like the iron, these impurities trickle to the bottom of the furnace; however, because they are lighter than the molten iron, the impurities remain on top and are called *slag*. Since the slag is made up of impurities, it is not wanted and is drawn out of the furnace. Most of the slag is removed in this way, but some remains and combines with the molten iron. It is this slag which later forms some of the discontinuities found in metals.

As the molten iron is drawn from the blast furnace, it is poured into molds to form what is called *pig iron*. This name came from an early process in which molten iron was poured into large molds called *sows*. These sows allowed the molten metal to trickle into smaller molds resembling suckling pigs, hence the name *pig iron*. Like its namesake, pig iron is not too clean. As the pig iron hardens, the slag impurities also harden into slate-like pockets within the metal. The impurities are nonmetallic and are accidentally included in the iron. These pockets of slag in the iron are called *nonmetallic inclusions*.

Pig iron is the first product in the steel-making process. It is too brittle for most purposes, so it is processed in an oven hearth furnace, along with other materials, to make better quality metal. When the ingot solidifies with many of the discontinuities contained in the upper portion, the "hot top" is cut off, also known as *cropping the top*. Note that not all discontinuities can be eliminated by cropping the ingot.

After the ingot is cropped, it is given a new name—*bloom*—that in turn can be processed into smaller size *slabs* or *billets*. The slab or billet is normally the starting point for the actual forming of articles or materials.

Rolling of the billet is typically done after the billet has been heat soaked in a furnace so that it is evenly heated throughout. Heating allows the crystals to break more easily into smaller grain-shaped crystals in the metal.

After heat soaking, to attain a uniform temperature, the billet is forced between large, heavy rollers. This rolling reduces the thickness of the billet and increases its length. Steel is rolled to reduce it in size and to shape it as close as possible to the shape of the finished product.

When a slab is rolled, the grain is formed in the direction of roll when the crystals are broken and reformed into small crystals. If the slab is rolled into sheet or plate material, it is rolled between wide, heavy rollers. Porosity and nonmetallic inclusions can cause a lamination. The gas porosity or inclusions in the slab will spread out in all directions, but mainly in the direction of roll.

Forging discontinuities

Forging is the working of metal into a desired shape by hammering or pressing the metal while it is very hot and in a soft condition. Since the forging process causes some discontinuities itself, the forging process starts with forging stock (alloyed for properties desired). The forging is then heated to the forming temperature and hammered into shape.

Discontinuities are typically connected to the process and temperature when the forging and the hammering process take place. The two conditions discussed below are caused by improper heat during the forging process. Die alignment is another common cause of the forging lap when the dies were not aligned, as illustrated in Figure 4.22; a forging lap at the seams would be produced.

Casting discontinuities

Castings are made by pouring liquid metal into a mold. These molds are formed close to the shape of the finished part. For example, many people cast their own bullets by melting lead and pouring it into a mold. When the metal solidifies (hardens from liquid metal to solid metal), it is removed from the mold.

Casting molds are typically made from sand, clay, and water. The clay and water form a thin film over each granule of sand, joining the granules to make the mold. As you might suppose, the mold can be easily broken and is permeable (absorbent). These are the desired qualities for a mold. You might also have guessed that castings have no regular grain structure. There is no rolling or forging to give direction to the grain. Figure 4.23 shows a casing being poured. When it solidifies, it will have an irregular grain structure.

Forging Discontinuities

Forging Discontinuities and Cause

• Forging Lap—Alignment of Dies

• Internal/External Burst or Crack—
Improper Temperatures, Not Hot Enough

Internal Burst
(Subsurface)

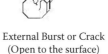

External Burst or Crack
(Open to the surface)

Figure 4.22   The origin of laps, internal and external bursts.

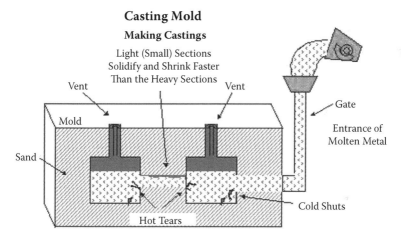

Casting Mold

Making Castings

Light (Small) Sections
Solidify and Shrink Faster
Than the Heavy Sections

Vent   Vent

Gate

Mold

Entrance of
Molten Metal

Sand

Cold Shuts

Hot Tears

Cold Shuts — Open to the surface, caused by cold spatter
and hot molten metal

Hot Tears — Tears of cracks caused by uneven shrinkage
during solidification

Figure 4.23   The origin of cold shuts and hot tear/cracks.

Some common types of discontinuities found in castings are cold shuts
and hot tears.

• *Cold shuts* are caused when molten metal is poured over solidified
  metal. When the metal is poured, it hits the mold too hard and spatters
  small drops of metal. When these drops of metal hit higher up on the

mold, they stick and solidify. When the molten metal reaches and covers the solidified drops of metal, a crack-like discontinuity is formed.

- *Hot tears* occur in a casting having light and heavy sections. The light section, being smaller, solidifies and shrinks faster, pulling the heavier sections, which are hotter and do not shrink as quickly, toward them. This unequal solidifying causes the faster shrinking light sections to apply a pulling stress on the heavier sections. This creates stresses, which can result in hot tears at the junction of the light and heavy sections. Hot tears can be prevented if, when stresses are built up by unequal shrinkage, the mold breaks before the metal tears. This breaking of the mold as the heavier sections are pulled against it takes up the stress that might otherwise tear the metal.

Verification points: receiving, first article, in process, and final

In order for a manager to effectively control a project, he or she must understand accountability, identification control, objective evidence, planning processes, and the complexity of the project.

- *Control is always present.* Although certain kinds of control are limited to certain phases during the manufacturing processes, management control of the process is sometimes as simple as casual observation, and is needed in every stage of operations for every kind of product or service. It is customary in many organizations to label, in general terms, the management control procedures according to the state of the product being examined. Some labels are *receiving, in-process, first-article*, and *final inspections*.
- *Receiving control.* The term *receiving control* denotes all the controls, regardless of type, that are given to incoming material, including such things as raw materials, specialty items, and subassemblies manufactured under subcontract work. To cut down on transportation and handling, companies making use of large quantities of specialty items or subcontract work frequently perform this kind of control in the supplier's plant.
- *In-process control.* Control that is conducted during the time raw materials are being converted into a finished product is called *in-process control*. The place of control is dependent largely upon the degree of examination and the kind of equipment needed. When only a percentage of the parts produced are inspected, either periodically or in spot checks, the work is usually carried on at the machine or production line. Particularly in small plants, the operator him or herself may perform this control. When large quantities of product are to be inspected,

and when the control procedures require specialized equipment, the work is most often done in centralized areas.

- *First-article control part of in-process or final control.* Regardless of the amount of other controls that might be necessary, *first-article control* is a common practice in a production line of more than one. After any equipment setup, tool change, personnel, or action that may influence the quality of the product, the first piece is examined to determine its conformance to manufacturing specifications. This is sometimes a very formal procedure, and in many cases, as in press punch work where the effect of wear and other factors is small, this may be the only control required.
- *Final control.* Control performed at *completion of the project* may include a great variety of work. Visual verification for appearance (finish, labels, cleanliness, and completeness along with all parts, all instruction books, a parts list, etc.) is nearly always part of the job. Tests for function, which are sometimes necessary on mechanical goods, may involve elaborate testing procedures requiring much time and adding considerable cost to the overall manufacturing operation. Testing of most aircraft and submarines in the final stages falls into this category. When the amount of final control is large, reduced in-process control may be permitted, although this will depend on a number of factors, including the relation of the control cost to processing cost and the cost of replacing bad parts in the final assembly.

Nondestructive testing

Nondestructive testing (NDT) is one of my greatest passions, and has held great importance for me during my career.

When I entered the U.S. Navy in 1963, I was sent to a nine-week training course on pipefitting and sheet metal work at the Naval Training Center in San Diego, California. During the course, I was first trained in how to mathematically lay out piping systems and sheet metal ventilation systems, then I was given practical exercises and a written and practical test.

What I realized at that time was that solving math problems came easily to me, and was an invaluable tool for my nondestructive testing background. Later in 1963, I attended a 19-week course of NDT training. The first phase was a four-week course on visual inspection, and the use of standards and specifications to determine requirements, methods, and acceptance criteria. Candidates with unsatisfactory marks in the first phase were dropped from the course. The second phase comprised the nondestructive testing methods.

There were two certification levels, and two categories of inspectors: NDT for Nuclear Reactors and Non-Nuclear NDT. The categories were visual (VT), magnetic particle (MT), dye penetrant (PT), radiography (RT), and ultrasonic (UT) inspector. I was one of the few who made it through the course to be certified as an inspector in Nuclear and Non-Nuclear VT, MT, PT, RT, and UT.

During my first tour of duty as an NDT inspector, I faced the reality of life as an inspector. My first mentor was a first class petty officer (Don) who was feared and despised by the crew of the repair department. I had found out that Don was one of the quality assurance inspectors for the repairs and overhaul of the USS *Thresher* SSN 593. The USS *Thresher* sunk with 112 shipmates and 17 fellow civilian workers. She sank on April 10, 1963. When I first reported, Don asked me what inspection processes I was qualified to do. I told him I was qualified across the board as an inspector in both nuclear and nonnuclear systems. He next asked me what shipyard or command I had come from. I told him I was directly out of school. He looked at me with a disgusted look and showed me the X-ray developing room; he then said to start cleaning, and don't stop until I have the time to start training you. It was one week before he started training me. His first and repeated lesson for three years was to never let anyone convince you to ever allow a local waiver, or allow a deviation from the specification or standard to take place. I had the opportunity, about a year later, to be with Don on an occasion where he cried like a baby over a couple of beers, and I felt his deep sorrow and the connection that he had with his inspector friends who had allowed the weaver to acceptance requirements that may have been the root cause that killed the crew and yard worker aboard the USS *Thresher*.

As you can guess from its title, *nondestructive testing* is defined as inspections that will not damage the material under test. While some processes may be detrimental to the material, or have a long-term damaging effect on the parts, components, or systems under test, when used properly, the chance of damage occurring is small.

The potential for damage is connected primarily with the magnetic particle and dye penetrant. There are various types of MT inspections, including wet and dry methods. The wet method is usually used on machined surfaces. MT methods can also be direct or indirect, meaning that the current is passed through the material under test, or the material under test is subject to a magnetic field. PT comes in water-washable and oil-based and wet and dry developer methods; oil-based methods are not used with the dye penetrant because it is corrosive and can get trapped in crevices. Dye penetrant can be used on any material that is not porous, such as glass and composite materials, if there are no detrimental effects.

Let's begin with the two that are together, MT and PT. Most standards call for an MT-PT inspection. The method used is based on variables that are required to make a calculated decision.

Magnetic particle inspection

- Magnetic particle inspection can only be performed on ferromagnetic materials, such as materials with iron that can be magnetized.
- It cannot be used on nonferrous materials such as aluminum, copper, 300 series stainless steels, and copper nickel.

- MT can detect subsurface discontinuities up to 1/4 inch beneath the surface.
- MT can induce a permanent magnetic field within the material that can interfere in other processes, such as welding and machining.
- When using the direct method (inducing current through the material under test), heavy arc strikes can occur similar to a welding rod strike. This localized concentration of heat needs to be removed through grinding and reinspection using the indirect method for surface cracks. Arc strikes are not wanted on machined surfaces.
- A fine metal dust is sprinkled over the part to be tested; this dust is a breathing hazard, and protective breathing masks must be used.
- This fine metal dust is ferromagnetic with high permeability and low retention. This means that the dust can be easily attracted to a magnetic field, but does not retain the magnetism.
- Parts under test need to be cleaned down to the bare metal. This means no paint, oil, film, rust, dirt, and the like.

How magnetic particle inspection works

The theory is based on the fact that magnetic metals attract. The field in the magnet that attracts metal is the magnetic field outside the magnet, or what is known as the *flux leakage*. If there is no flux leakage, the material will not act like a magnet. It is important to note that it is possible for a material with an internal magnetic field to not have flux leakage, and therefore be unable to attract metal.

An example of the magnetic particle inspection process, illustrated in Figure 4.24, can occur when a calculated current is passed through a part at an amperage of about 100 to 125 amps per prod spacing. If there were no interruptions in the induced current the MT dust puffed onto the area under test, it could drop to the surface like baby powder. If there is a brake or interruption in the induced current, the metal dust will be attracted to the magnetic +/− of flux leakage. The flux leakage will have a positive and negative effect and will attract the MT dust, usually in the shape of the interruption. An internal crack will have a dust liner shape that is more than three times its width, indicating a liner composed of metal dust. Porosity will appear to be a round or elliptical accumulation of metal dust.

Dye penetrant

Dye penetrant is used to detect only surface indications of flaws. It cannot detect subsurface flaws. The method used in dye penetrant is based on variables that must also be used in making calculated decisions.

Dye Penetrant Inspection (PT)

- PT can only detect discontinuities open to the surface.

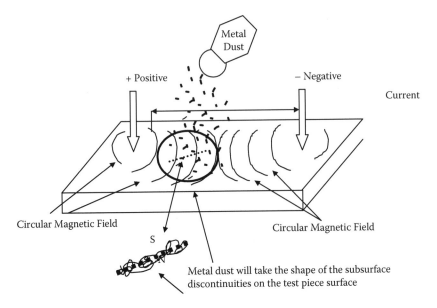

Figure 4.24   The magnetic particle (MT) inspection process.

- It can be used on all nonporous materials; a porous material like ceramic cannot be inspected using PT.
- PT in the oil base form can be corrosive to metal if it gets trapped in crevices such as bolthole threads and not cleaned off.
- PT is easy to transport to the area under test because it typically is available in aerosol cans, or paint cans.
- The dye is wetter than water (note: dye penetrant has a much lower viscosity than water and combined with low surface tension can enter tight cracks that water cannot enter) and can float in the air, so breathing protection is required.
- For fluorescent PT, a black light and a dark area for testing should be made available.
- There are three components to the kit: dye, cleaner, and developer. Other items can include acetone, lint-free wipes, and a black light if using the fluorescent method.
- PT is a breathing hazard in confined spaces; the inspector should have good ventilation and wear a protective mask.

How dye penetrant inspection works

When using the dye penetrant inspection method, illustrated in Figure 4.25, the test surface must not have any paint, dirt, lint, grease, oil, or the like on it. The dye penetrant comes in four groups; we will be using group 1, a solvent removable visible dye, in this example. The dye penetrant (bright red and wetter than water) is typically brushed evenly on the test surface. Spraying

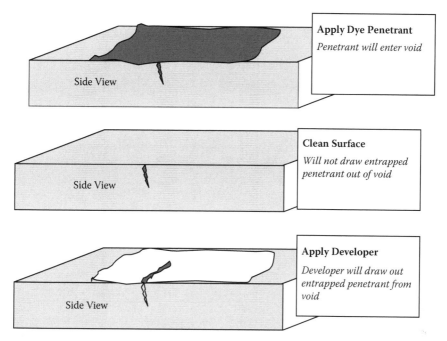

Figure 4.25 The dye penetrant (PT) inspection process.

the surface can be messy; instead, the spray can cap is usually filled with dye, and then it is brushed onto the testing surface. Group 1 has a dwell time (time kept wet on the part) of about 15 to 20 minutes. At the end of the dwell time, the area is first wiped off with a dry wipe, and then slightly dampened using the cleaner until it turns a slightly pink color. Next, the surface is lightly sprayed with developer to produce a white surface. The inspector then watches the surface and waits. The interpretation must take place a minimum of seven minutes after the developer is applied, and no later than 30 minutes after.

Radiographic inspection

X-rays and light are forms of electromagnetic radiation, and are used in radiographic inspection. Both inspection methods are dangerous to the radiographic operator as well as any personnel who may be in the area. When radiographic operations are performed, all personnel in the area must be notified so that the area can be cleared. Also, an emergency action plan must be filed with the radiographic officer before the start of radiography procedures.

X-rays differ from light due to their extremely short wavelength; X-rays are only about one ten-thousandth or less the length of light. This gives X-rays the ability to penetrate materials that absorb or reflect ordinary light. X-rays exhibit all the properties of light, but they modify their particle behavior.

Radiographic inspection uses X-rays (manmade) or gamma rays that are similar, but differ in source because a machine produces X-rays, and isotopes produce gamma rays. Some isotopes are gamma-ray-emitting isotopes such as radium, which occur naturally in nature. Others, like iridium 192 and cobalt 60, are artificially produced. In industrial radiography, the artificial radioactive isotopes are used almost exclusively as sources of gamma radiation. The electromagnetic spectrum chart illustrated in Figure 4.26, provides a view of the light spectrum in relation to visible light.

X-ray and gamma ray radiography uses the same process as a doctor's office during a chest examination. The source produces an X-ray or gamma ray that passes through the patient's body, and the doctor interprets the results. Figure 4.27 illustrates the interpretation of the film or computer screen that is based on the amount of radiation that develops the film. The fewer the rays, the lighter the image is; the more the rays, the darker the film is. This means that a mass of bone will be lighter, and if there is a break in the bone, it would appear as a dark line in the shape of the break. A dense mass such as cancer growth will appear lighter than the surrounding area. The same principles apply to industrial radiographs. Gas pockets have less metal, so they will appear darker and round or elliptical. Cracks have less material so they will also be dark, but will appear linear and jagged. Slag from welding will also be dark. Tungsten inclusions for the GTAW process of welding will be lighter because tungsten is denser than steel, meaning less rays will hit the film or receiving plate connected to the computer.

Ultrasonic inspection

I have witnessed incredible advances in ultrasonic inspection (UT) during my career, first in the industrial arena, and then moving into the medical arena.

In the early 1960s, ultrasonic equipment was basically a piezoelectric transducer and an oscilloscope. All the calculations and calibration of the

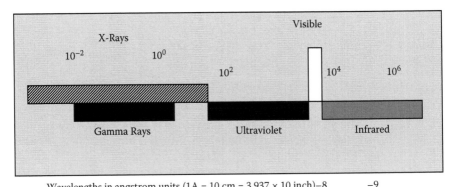

Wavelengths in angstrom units (1A = 10 cm = 3.937 × 10 inch)–8   –9

Figure 4.26 Electromagnetic spectrum chart.

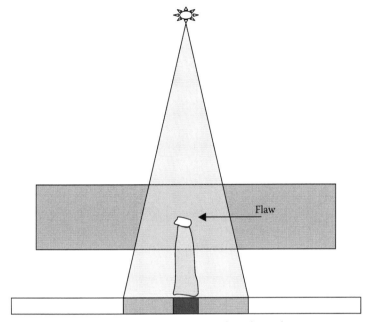

Schematic Diagram Showing The Fundamentals of a Radiographic Exposure

Figure 4.27   The interpretation of the film or computer screen is based on the amount of radiation that develops the film.

unit were dependent on the ability of the inspector. During the inspection process, it was like being in a cave viewing stalactites, and any spikes seen by the layperson would be actual discontinuities and defects.

When an ultrasonic sound wave is sent through the metal, the results are based on the measurement of time that it takes for the ultrasonic sound to leave the transducer and the amount of ultrasonic sound that returns. The transducer is a piezoelectric crystal; the term *piezoelectric* means to change electrical energy into mechanical energy (ultrasonic sound), and then convert the returned sound to electrical energy. A "gate" in the equipment pulses to allow the giving and receiving of sound, and the electrical conversion in the unit converts the electrical energy into the time and amount of signal returned.

Illustrated in Figure 4.28a is the longitudinal (straight-beam) sound wave and the shear wave (angle-beam) methods of inspection (Figure 4.28b). Discontinuities lay at all different angles to the surface; therefore, by providing different directions of sound, discontinuities can be picked up at different angles to the surface. Shear wave is popular with weld inspection of full-penetration welds because the end prep on full-penetration welds is beveled, usually on a 45° angle or something close to 45° (Figure 4.28c). As you can see from Figure 4.28b and c, the shear wave would be perfect to inspect the heat-affected zone (HAZ). And, as you remember, the HAZ is the most critical inspection area.

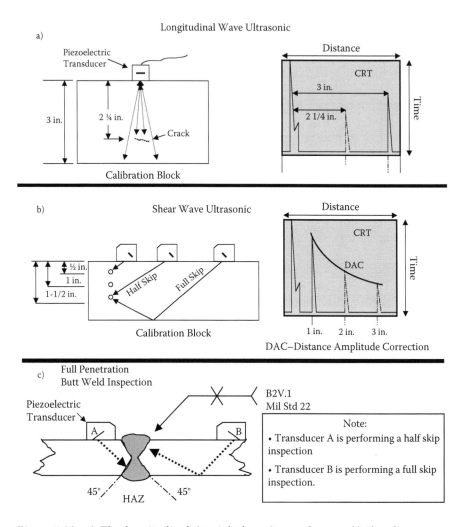

Figure 4.28 a) The longitudinal (straight-beam) sound wave, b) the shear wave (angle-beam), and c) the full penetration Butt weld.

There are also other types of NDT inspections, such as an eddy current, but the most important inspection is the first inspection performed, which is the *visual* inspection of the preparation and construction of materials and parts, the items in process inspections, and final inspections before any NDT is performed. If the visual inspections are not performed correctly and with wisdom and knowledge, the NDT inspection may have less of a chance to find the discontinuity that could cause the part to fail in service.

Destructive testing

Destructive testing is a process that subjects a product or material to tests that simulate critical life cycle situations. The intent is to create a controlled environment that creates situations that could be detrimental to the user, the environment, or the safety of others. One example of this is the cabin atmosphere in a passenger airplane at 30,000 feet. Airline passengers board a plane trusting that the air is breathable and not contaminated, yet there are certain situations that can jeopardize the safety of the aircraft passengers, including the presence of toxic gases from an electrical fire or the loss of cabin pressure from a window breaking or a wing falling off. It is obvious that it would make sense to test the aircraft on the ground in a controlled situation to determine the structural soundness of the material in the wing, ensure that the windows will not break, and make sure that the passengers would not die from toxic gases in the event of an electrical fire.

There are many types of destructive tests available in the aerospace industry, as well as new ones being developed, that ensure that equipment meets safety standards. In this section, I will explain a few destructive test methods in order to give you a foundation of the principles that are used.

Tensile tests

Tensile testing, illustrated in Figure 4.29, is used to test many materials for strength and toughness. Although tensile testing is done on fabric, glass,

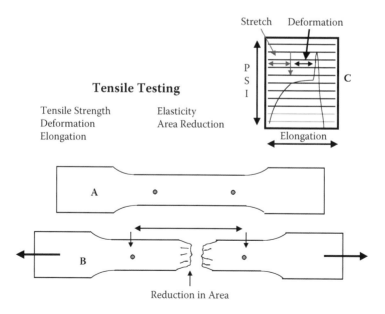

Figure 4.29 Tensile testing is used to test many materials for strength and toughness.

plastic, and many other products, we will discuss its application in metal, and use an aircraft wing as our example. In order to determine the safety of the aircraft wing, it is important that we understand the characteristics of the metal that is used to manufacture it, including its toughness and strength. Some of the useful characteristics of metal that need to be known by the engineers designing the aircraft wing are the tensile strength, elasticity, and breaking point in pounds per square inch.

Fracture toughness tests

There are two main types of fracture toughness tests: Charpy V-notch and fracture toughness. Both methods are defined by the American Society for Testing and Materials (ASTM) G 129-00 standard, titled *Standard Practice for Slow Strain Rate Testing to Evaluate the Susceptibility of Metallic Materials to Environmentally Assisted Cracking* (ASTM 2000).

The first test to be developed was the Charpy V-notch (Figure 4.30, view A-1). The test is primitive and has remained the same for centuries. As part of the test, a hammer is released and allowed to swing and strike the V-notch sample (Figure 4.30, view A-2); the amount of travel past the sample is measured to determine the amount of resistance to fracture the sample. Some tests call for elevated- (high-) temperature samples, ambient (room) temperature, or subtemperatures (about –150°C). The test is simple and requires only a standard test fixture that can be made by the test facility. The test basically measures the amount of energy to fracture a sample of material.

The fracture toughness specimen is used to measure the resistance of a sample of material to resist a vibration and strain. The equipment used in

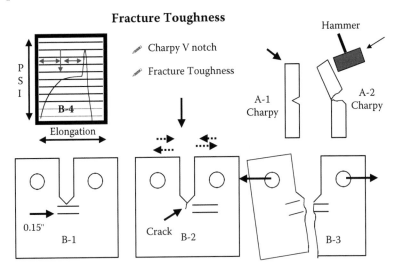

Figure 4.30   Fracture toughness tests.

this test is typically more sophisticated and involves a computer drive cycle strain machine. The equipment is attached to the two holes in the specimen, and two lines are scribed on the specimen, one at 0.05" and the other at 0.15" from the bottom of the V-notch (Figure 4.30, view B-1), then the specimen is slowly subject to known cycles (vibration) and tension (Figure 4.30, view B-2). The V-notch is observed, and the length of the crack (Figure 4.30, view B-3) that appears is recorded when it reaches the 0.5" and 0.15" mark; the test is then stopped, and the cycles and tension are recorded. The next part of the test will record the amount of tension to rip the specimen apart. This test is also performed at other than ambient temperature to fit the environment and stress the part is subject to in service.

Both tests are typically performed twice, with one sample taken with the grain flow of the material lengthwise, and the other taken at 90° across the grain flow. This is important to give a comparison of strengths.

Prework review

When a quote for a bid, rework, or a new project is received, it is up to the organization to determine if the attributes have changed. These attributes can be concept changes, drawing changes, or modifications to the original contract description, delivery date, long-lead materials required to accomplish the project, or cost of special instructions or specific suppliers. The organization must also determine how much these amendments will cost. Additionally, the customer's purchase order or contract requirements must be complete and not ambiguous.

For example, if the customer asks for a zinc-plated spring with no specifications, this indicates that the spring will be in a corrosive environment. If the spring is sent to a processing house for plating, it could be flash-coated with a thin zinc finish that may cause rusting of the spring in the customer's storage before installation. Who is going to be responsible for this? The customer or you?

Product realization for the customer-related processes is understanding the bigger picture, not just capturing the customer requirements. You must take the time to understand your customer's needs:

1. What is the customer's product used for?
2. Can we get closer to the customer through the product?
3. Can we pick the specifications for the customer?
4. Has the customer by virtue of his function dictated the specifications?
5. Can we deliver on time and make a profit?

Sometimes what the customer does not put on the purchase order can cause a much larger problem. While the customer's engineering and quality departments are required to add all engineering and quality clauses and specifications, the customer's purchasing agent may not include them. Who is responsible? Who will pick up the bill for wrong products or services? Because you are considered the professional in your area of expertise, you will most likely either pay the entire cost or share the cost to make it right!

Drawing design, perception, and execution

Introduction and background to engineering drawings

The standard practice for engineering drawings, MIL-STD 100G, was cancelled on June 9, 1997. Future requirements for engineering drawings should refer to the following American Society of Mechanical Engineers (ASME) standards, as applicable: ASME Y14.100, Engineering Drawing Practices; ASME Y14.24, Types and Application of Engineering Drawings; ASME Y14.35M, Revision of Engineering; and ASME Y14.34M, Associated Lists (ASME 2000).

The U.S. government was a major contributor of research and development dating back to World War I. By World War II, the United States was heavily involved in the development of military standards and specifications that captured the trials and errors of providing products or services. The U.S. Department of Defense (DoD) and military standards (MIL-STD) defined the materials, fabrication requirements, assembly instructions, workmanship, and acceptance requirements for the United States and its allies.

The common practice to reduce costs in commercial and governmental organizations has had a latent effect on the quality of products and services used, and the loss of financial control of projects.

One of the most successful implementations of this practice is in the Navy Nuclear Program, Navy Subsafe Program, Boeing Aerospace C-17 project. Their drawings have been continuously improved to present clear instructions, with few ambiguous or missing requirements for completing the task. The success of these programs is based on the persistence and discipline of the product quality-planning (PQP) team. They may have different motivations, but the end result produced very successful programs.

To use an understood common practice is important. Some of the key drawing requirements are listed below.

ASME Y14.100-2000 Practices: USS Dolphin Training Points

- Training point 1: flag-note symbols used on the drawings:
- Training point 2: commonly used words and phrases; the following rules should be applied:
  - Reference documents shall be cited using "per," "conforming to," "as specified in," and "in accordance with" (or "IAW").
  - The phrase "unless otherwise specified" shall be used to indicate the generally applied requirements, and should appear at the beginning of the note or denoted at the head of the "Notes" column.

Table 4.3   Acronyms for Special Items and Processes

Acronym	Description
CSI	Critical safety item
CSP	Critical safety process
ENI	Environmental impact
ESD	Electrostatic discharge sensitive devices
ESS	Environmental stress screening
HAZ	Hazardous conditions, processes, or material
HCI	Hardness critical item
HCP	Hardness critical process
I/R	Interchangeability/reparability
INT	Interface control
ODC	Ozone-depleting chemical
ODS	Ozone-depleting substance

Source: From ASME (2000).

> This phrase shall be used when providing a reference to another document, or a requirement on the drawing, that clearly specifics the exception(s).

- Training point 3: indefinite terms; indefinite terms such as *and/or, etc.,* and *e.g.* shall not be used.
- Training point 4: language style; notes shall be concise statements using the simplest words and phrases for conveying the intended meaning.
- Training point 5: illustrated in Table 4.3 are acronyms for special items and processes.

The tailoring checklist

ASME.Y14.100-2000 (ASME 2000) shall be carefully tailored to meet the user's requirements. It is essential that the applicability of the numerous referenced documents, especially regarding basic practices, be as definitive as possible. The tailoring checklist in Table 4.4 is used to define who has the authority and responsibility to determine key attributes, such as how to actually develop the design and control the configuration. Table 4.4 demonstrates that the drawing format offers a great deal of configuration management control. Liability is typically the main issue that develops during the design process. Liability determines who will pay for the design problem during the life cycle. However, liability cannot be assigned without defined controls over the form, safety, fit, and function. The tailoring checklist in Table 4.4 lists the decisions that must be made. It is important to remember who is liable, or responsible, for mistakes during the design process.

Table 4.4   The Tailoring Checklist

A. Drawing Media (Choose All That Apply)

1. Nondigital data (specify......)
2. Digital data (specify......)
3. Other (specify......)

B. Drawing Format (Choose One)

1. Contractor
2. Government (forms supplied by the government)
3. Government (forms supplied by the contractor)

C. Drawing Sheet Size and Format (Choose One)

1. ASME Y14.1
2. ASME Y14.1M

D. Application Data (Choose All That Apply)

1. Contractor option
2. Required
   a. On drawing
   b. By reference (specify......)
   c. Contractor option
3. General use or multiuse notations
   a. Allowed
   b. Not allowed

E. Drawing Detail (ASME Y14.24; Choose All That Apply)

1. Monodetail
2. Multidetail
3. Tabulated

F. Dimensioning and Tolerance (Choose All That Apply)

1. Metric
2. Decimal-inch
3. Application of ASME Y14.5M
   a. Specific issue (rev.) required (specify issue......)
   b. Issue in effect (specify issue......)

G. Drawing Notes (Choose One)

1. On drawing
2. By reference (specify......)
3. Contractor's option

H. Types of Drawings (ASME Y14.24; Choose One)

1. Contractor selects
2. Government selects

Table 4.4   The Tailoring Checklist

I. Maintenance of Multisheet Drawings (ASME Y14.35M; Choose All That Apply)

1. Drawing revision level (DoD preferred)

2. All sheets same revision level

3. Sheet revision level

   J. Redrawn Drawings (Redrawing without Change; ASME Y14.35M; Choose One)

1. Advance revision level

2. Revision level is not advanced

   K. Maintenance of Revision History (Choose All That Apply)

1. Contractor's option

2. Optional method

   a. Remove one or more revision record(s) as required

   b. Remove all previous revision history

   c. Remove all revision history but retain line entry for revision authorization and date of revision

   d. Remove all except revision preceding current

   e. Maintain revision history in its entirety

   L. Adding Sheets (ASME Y14.35M; Choose All That Apply)

1. Contractor's option

2. Optional methods

   a. Renumber sheets using consecutive whole numbers

   b. Number added sheets in decimal-number sequence

   c. Number added sheets in alphanumeric sequence

   M. Deleting Sheets (ASME Y14.35; Choose All That Apply)

1. Contractor's option

2. Optional methods

   a. Renumber all affected remaining sheets

   b. Affected remaining sheets not renumbered (revision status of sheets block is updated with notations such as CANC [cancelled] or DEL [deleted])

   N. Markings on Engineering Drawings (Choose One)

1. Special items and processes apply

   a. Applicable symbols (specify……)

   b. Applicable special notes (specify……)

2. Special items and processes do not apply

   O. Associated Lists (ASME Y14.34M; Choose All That Apply)

1. Nondigital data (specify……)

2. Digital data (specify……)

3. Other (specify……)

Table 4.4   The Tailoring Checklist

P. Types of Associated Lists (ASME Y14.35M; Choose All That Apply)

1. Parts lists
   a. Integral
   b. Separate
   c. Contractor's option
2. Application list
3. Data lists
4. Index lists
5. Indentured data list
6. Wire list
7. Other (specify......)

Q. Angle of Project (Choose One)

1. Third angle
2. First angle

R. Language (Choose One)

1. English required
2. Other (specify)

S. Applicability of Appendices

1. Appendix B
   a. As detailed herein
   b. As modified
2. Appendix C
   a. As detailed herein
   b. As modified
3. Appendix D
   a. As detailed herein
   b. As modified
4. Appendix E
   a. As detailed herein
   b. As modified

Source: From ASME (2000).

Interpretation of fabrication standards and specifications

(*Note*: This next section is crucial to maintaining a successful career in any sector of the workforce.)

Standards and specifications do not provide an enjoyable reading experience. There are no metaphors or similes as in poetry. There is no mayhem or suspense like you would find in a good mystery novel; no jokes as in a comedy; and, worse yet, no plot or climatic ending. Standards and specifications may, however, make good bedside reading because they can put you to sleep if you do not understand the author's intent. Let's examine the initial intent of military standards and specifications, because that is where it all started.

Standards and specifications first began to appear in the U.S. Navy as it was in the process of building a fleet of ships. The Navy required that all the sailors have the ability to repair the ships while out to sea, as well as after battle. This meant that all Navy personnel would have to be able to identify every material, its method of fabrication, and the operating and maintenance instructions for the systems. As a result, it became crucial that all specifications were standardized throughout the Navy.

The architects of the original Navy standards and specifications had to create documents that illustrated exactly how a ship was to be built in language that could be understood by sailors and other ship personnel, on an eighth-grade reading level.

A perfect example is the NAVSEA 250-1500-1 standard (U.S. Navy n.d.), for which I was able to have access to the authors. As some of you may know, the reactor plant manuals (RPMs) defined the complete map of how to operate, maintain, and repair nuclear and associated systems.

Admiral Hyman Rickover was the driving force behind NAVSEA 250-1500-1, and demanded communication and support throughout the entire nuclear program. The NAVSEA 250-1500-1 was referred to as the fabrication and repair document for the mechanical systems. It is a complete standard, and it includes requirements for materials, fabrication, personnel qualifications, inspection requirements, acceptance requirements, repair requirements, and much more.

This standard is known as the "Bible" of work performed within the reactor system, including two valves or boundaries outside the system. My contact was a man named Bruno; he along with others helped in the understanding and interpretation of the standard. When I would ask a question about how to apply a standard requirement, he would give me the reason that it was written the way it was, and how to apply it. It was almost like giving me a parable so I would always look for the least obvious answer.

In one situation I was confused on a few items; one was what I considered a minor requirement that had a heavy emphasis in the standard. The requirement stated there shall be no undercut in excess of 1/64 of an inch allowed. In addition, there was another requirement stating that any abrupt edge would have to be contoured to meet surface requirements. I did not see the harm in a little undercut, but I discovered later that I was wrong.

When Bruno explained how they came up with the requirement, he was able to answer hundreds of related questions with a single answer. He told me that the reason for the strict requirements was based on the tests that had been performed to determine the severity of all types of discontinuities in welding. They first made a large steel tank and included every known type of discontinuity they could think of.

During the test, there were known amounts of internal inclusions, voids, cracks, and surface conditions such as undercut, and excess and not enough weld reinforcement. They filled the tank with water, and then started to drop procession grenades into the tank. At predetermined times they would perform nondestructive testing such as X-ray, ultrasonic, magnetic particle, and dry reentrant tests and map the propagation (growth) of the discontinuities.

Bruno explained the results and gave me copies of the reports. To my surprise, the undercut was at the top of the list. Internal discontinuities such as cracks and other defects not typical to the surface were the least slow or fastest to propagate. The reason for the undercut was not only its abrupt change in section, but also its location in the heat-affected zone.

As you can probably guess, the HAZ was a thin area at the boundary of the weld joint that is half melted. Because the molten (liquid) electrode and base metal flow into each other, the thin HAZ area is not fully melted, and not fully solidified (solid metal), thus putting it in a questionable category that cannot be reasonably predictable. As an example, when we weld a piece of 304 stainless metal to another piece of 304 stainless metal, we use 308 filler metal. The 308 filler metal when used with a qualified (proven) weld procedure and qualified welder (person using the qualified procedure) theoretically will produce a weld that has the same strength and properties as the 304 base metal. As you can imagine, the HAZ is the only area in question, and has proven to be the weakest link of the process. Thus, any slight abrupt change (such as undercut in excess of 1/64 inch) in the HAZ surface will propagate over time, causing a crack and possible catastrophic failure. This is just one of many examples I can give you to assure you that standards and specification must not be taken lightly. Even the engineer must beware of changes made that are not in line with these documents and are outside the scope of applied standards.

The standards and specifications are meant to give all the engineering information that was accumulated during the fabrication of the ship to the ship. Each ship has the capability through sailor training to repair just about any problem. The ships have materials on board to do it also. An example is the submarine, which has the high-quality materials cataloged throughout the ship; these are used in a practical way but are available if necessary for emergency purposes, such as some level-1 piping that is used throughout the boat. What seems like an ordinary handrail on a stairway may be the 1-inch, level-1 copper nickel piping needed in the event that the nuclear reactor coolant discharge system needs to replace a section of pipe in an emergency situation. As you can see, there is good reason to use the standards and specification to build or repair a ship. The simple substitution of material can be devastating if an emergency repair cannot be made at sea because the substituted material is not compatible with the repair materials on board, such as a compatible welding rod or an abnormal O-ring size.

The "sea lawyer"

When asked how I coped with 20 years in the high-stress Nuclear Subsafe Submarine Navy repair, modification, and overhaul sector, I replied that I was a "sea lawyer." I was on a newly commissioned submarine tender in San Diego, and I remembered being in the Navy before I realized the sea lawyer method of surviving. I felt that I had no say in what was going on, and marveled at the knowledge that some people possessed—until I was given a well-kept secret by a lieutenant commander who I had an argument with. This changed my life forever. As the division commander was escorting the lieutenant commander out of the Nondestructive Testing Laboratory on the ship after we had argued, he threatened to get even with me. About two weeks later, I received a call to report to his office. I remember saying to myself on the way down to his office, "This was payback time."

I stood in front of the lieutenant commander, and he proceeded to tell me how much he disliked me, and never would like me, but he needed my help. He said he was tasked to establish the Nuclear Systems Repair Department that could achieve certification to perform nuclear repairs within three months. He needed a leader and work coordinator who was street-smart intelligent and respected by the enlisted workers. I told him I did not believe it would be a good idea for me to work with him. He said that if I did work with him, he would teach me everything I wanted to know about nuclear submarines and support me 100 percent in my decisions. However, if I made a mistake, he would come down hard on me.

Shortly after I accepted his offer, his knowledge and mentoring were a light in my life, and one of the techniques he shared with me was the proper use and interpretation of standards and specifications. The *sea lawyer* was a title I gave myself early in my career, and later taught this valuable technique to students at the Naval Sea System Training Center for four years in the Navy, and throughout the rest of my career.

I found that by understanding standards, specifications, and the hierarchy of directives, I could accomplish what others could not. As in the law of the land, there was the law of the sea, and in the Navy the rules were established and orders were handed down from the president of the United States (commander-in-chief), to the secretary of the Navy, and all the way down to me.

The sea lawyer method can be used in corporate policies, manufacturing, fabrication, and even government agencies.

First, let's examine the fabrication of a product. The hierarchy must be established first; if we are talking about a corporation, we need to know the flow of directives from the board of directors' bylaws, to the CEO's directives, to departmental procedures regarding specific instructions for areas that need to be changed.

For the sake of staying focused on one sector, we'll use the government, because all the attributes are defined and consistent with those of the commercial industries. In fabricating a piping system for a ship's boiler:

- The commander-in-chief (president of the United States) sends a directive to the secretary of the Navy, such as, "In order to keep with the current times and because we have cancelled our Mil-Q-9858 Quality Management System used to define the organization and processes in the past, I want the military to implement an operating management system in line with the International Standardization of Organization ISO 9001:2008. As you know, the ISO 9001:2008 was based partly on the Mil-Q-9858 principles. It has proven to be effective in the aerospace industry. Make this requirement fully implemented by January 2008."
- The secretary of the Navy instruction (SECNAVINST) can be as follows: telling all repair, fabrication, and overhaul facilities to meet the guidelines of having a system of operation such as ISO 9001:2008. This system will maintain a configuration management of all items, systems, and services performed on ships and shore facilities that support the fleet. It will also call for a means of establishing a system to control the management of processes and functions in orderly, measurable means.
- The commands will focus the SECNAVINST to be tailored toward their specific command and mission, such as by issuing a directive naming the engineering group, supply group, facility maintenance group, and so on.
- The divisions within the command issue their specific mission statements, quality policies, and instructions that are all in line with the SECNAVINST and command instructions. They will address the policy, operating procedures, and instructions.
- The work groups will develop the specific work instructions to accomplish the work to be completed in line with all of the above, with the specific weight on the workmanship standards performed.

Organizational standard requirements

When we read and understand the intent of the hierarchy instructions in any organization, we will see that they are always looking for stability and control of what is going on.

If you put yourself in the top-man or top-woman position, you would do the same thing. This means that as the information flows down, the instructions and directives are altered through misinterpretation, and sometimes spun to meet the needs of an individual. When I decide to change something, I must know the intentions of the top directive. If it is in line with a situation I need to change or reinforces a view I have, I will work backwards (bottom-up) to stress my point of the change that needs to take place. Most likely, the problem is at the division or work group level.

If the division or work group does not have well-defined operation instructions, that's an easy one: they have to establish defined operation instructions. If I want to change a division policy, I use the higher directive as the basis of my argument. Why? The higher directives give a concept with an intention. My argument either could be to conform to the intent of the higher directive or (the other person or persons) to notify the higher directive authors, and explain to them that they are wrong and need to change the higher directive. Guess what: it isn't going to happen, and they will have no choice but to conform to your thinking.

Fabrication standard requirements

Fabrication standards are important because they contain all pre-engineered acceptable workmanship and fabrication requirements. We all know that engineers design and architect the system, but a nonengineer can determine if the engineer's drawing is correct in relationship to the fabrication document. For example, if I'm working on a superheated steam-piping system to run a turbine for the Navy, all I need to know is how much steam pressure, temperature, and volume are required, and I can verify the piping system and material of pipe, its criticality class, its acceptance criteria, the welding requirements and joint designs, and the inspection method and objective evidence required to document the completeness and acceptability of the installation.

The lead document is the bible of fabrication, which can solve any vague or ambiguous work instructions. Let's take two fabrication documents: Mil-Std 278 and AWS D1.1. Both documents define from start to finish all necessary requirements for fabricating a specific type of work. AWS D1.1 is a fabrication document in the commercial structural document, and Mil-Std 278 is for piping, machinery, pressure vessels, castings, and forging for Navy ships.

Both are similar documents because they give specific instructions for workmanship, welding, machining, materials, and much more. Both documents give you directly or reference specific acceptance criteria, requirements for fabrication, methods of inspection, workmanship requirements, and many other important fabrication requirements.

References

American National Standards Institute (ANSI). 1994. *ANSI/NCSL Z540-1-1994.* Washington, DC: ANSI.

American Society for Testing and Materials (ASTM). 2000. Standard practice for slow strain rate testing to evaluate the susceptibility of metallic materials to environmentally assisted cracking. West Conshohocken, PA: ASTM.

American Society of Mechanical Engineers (ASME). 2000. *ASME Y14.100-2000.* New York: ASME.

U.S. Navy, Naval Sea Systems Command (NAVSEA). N.d. *NAVSEA 250-1500-1: Welding standard.* Washington, DC: U.S. Navy.

Chapter five

Measuring instruments and the "lost art of the hand tool"

Over the years, the art of visual inspection and hand tools used for measurement has been diminished. Going as far back in time as Noah's ark, the lack of a yardstick was not a serious drawback.

During that era, most measuring was done by one craftsman completing one job at a time, rather than assembling a number of articles piecemeal to be assembled later. It didn't make much difference how accurate the measuring sticks were, or even how long they were.

Generally, the length of a mile, yard, or inch, or the weight of a pound or ounce does not make much difference. The cubit of Noah's time was the length of a man's forearm or the distance from the tip of the elbow to the end of his middle finger. Many times this was useful, because it was readily available, was convenient, and couldn't be mislaid. However, it was not a positive fixed dimension or a standard.

Following the rise and fall of the Roman Empire, about 600 years after the time of Christ, Europe drifted into the Dark Ages. For 600 or 700 years, humankind generally made little progress with regard to standardizing measurement. Sometime after the Magna Carta was signed in the 13th century, King Edward I of England took a step forward. He ordered a permanent measuring stick made of iron to serve as a master standard yardstick for the entire kingdom. This master yardstick was called the *iron ulna*, after the bone of the forearm, and it was standardized as the length of a yard, very close to the length of our present-day yard. King Edward realized that constancy and permanence were the keys to any standard.

Practically all shop jobs require measuring or gauging. You will most likely measure or gauge flat or round stock; the outside diameters of rods, shafts, or bolts; slots, grooves, and other openings; thread pitch and angle; spaces between surfaces; or angles and circles.

For some of these operations, you'll have a choice of which instrument to use, but in other instances you'll need a specific instrument. For example, when precision is not important, a simple rule or tape will be suitable. In other instances, when precision is important, you'll need a micrometer.

The term *gauge* or *micrometer* as used in this chapter identifies any device that can be used to determine the size or shape of an object. There is no significant difference between gauges and measuring instruments. They are both used to compare the size or shape of an object against a scale or fixed dimension. However, there is a distinction between measuring and gauging that is easily explained by an example. Suppose you are turning work on a lathe and want to know the diameter of the work. Take a micrometer, or perhaps an outside caliper; adjust its opening to the exact diameter of the work piece; and determine that dimension numerically. On the other hand, if you want to turn a piece of work down to a certain size without frequently taking time to measure it, set the caliper at a reading slightly greater than the final dimension; then, at intervals during turning operations, gauge, or "size," the work piece with the locked instrument. After you have reduced the work piece dimension to the dimension set on the instrument, you will, of course, need to measure the work as you finish it to the exact dimension.

Adjustable gauges

You can adjust adjustable gauges by moving the scale or by moving the gauging surface to the dimensions of the object being measured or gauged. For example, on a dial indicator, you can adjust the face to align the indicating hand with the zero point on the dial. On verniers, however, you move the measuring surface to the dimensions of the object being measured.

Micrometers

Micrometers are probably the most used precision measuring instruments in a machine shop. There are many different types, each designed to measure surfaces for various applications and configurations of work pieces. The degree of accuracy also varies, with the most common graduations ranging from one one-thousandth (0.001) of an inch to one ten-thousandth (0.0001) of an inch. I have provided brief descriptions of the more common types of micrometers in the following paragraphs.

- *Outside micrometer.* Outside micrometers (Figure 5.1) are used to measure the thickness or the outside diameter of parts. They are available in sizes ranging from 1 inch to about 96 inches in steps of 1 inch. The larger sizes normally come as a set with interchangeable anvils that provide a range of several inches. The anvils have an adjusting nut and a locking nut to allow you to set the micrometer with a micrometer standard. Regardless of the degree of accuracy designed into the micrometer, the skill applied by each individual is the primary factor in determining accuracy and reliability in measurements. Training and practice will make you proficient in using this tool.
- *Inside micrometer.* An inside micrometer is used to measure inside diameters or between parallel surfaces. They are available in sizes ranging from 0.200 inches to over 100 inches. The individual interchangeable extension rods that may be assembled to the micrometer head vary in size by 1 inch. A small sleeve or bushing, which is 0.500 inches long, is used with these rods in most inside micrometer sets to provide the

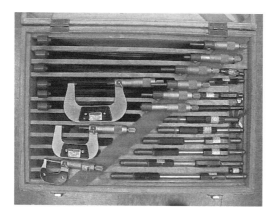

Figure 5.1 Outside, inside, depth, and thread micrometers.

complete range of sizes. It's slightly more difficult to use the inside micrometer than the outside micrometer. There is more of a chance that you won't get the same "feel" or measurement each time you check the same surface. The correct way to measure an inside diameter is to hold the micrometer in place with one hand as you feel for the maximum possible setting of the micrometer by rocking the extension rod from left to right and in and out of the hole. Adjust the micrometer to a slightly larger measurement after each series of rocking movements until you can no longer rock the rod from left to right. At that point, you should feel a very slight drag on the in-and-out movement. There are no specific guidelines on the number of positions within a hole that should be measured. If you are checking for taper, take measurements as far apart as possible within the hole. If you are checking for roundness or concentricity of a hole, take several measurements at different angular positions in the same area of the hole. You may take the reading directly from the inside micrometer head, or you may use an outside micrometer to measure the inside micrometer.

- *Depth micrometer.* A depth micrometer is used to measure the depth of holes, slots, counter bores, and recesses, and the distance from a surface to some recessed part. This type of micrometer is read exactly opposite from the method used to read an outside micrometer. The zero is located toward the closed end of the thimble. The measurement is read in reverse and increases in amount (depth) as the thimble moves toward the base of the instrument. The extension rods come either round or flat (blade-like) to permit measuring a narrow, deep recess or grooves.
- *Thread micrometer.* The thread micrometer is used to measure the depth of threads that have an included angle of 60°. The measurement obtained represents the pitch diameter of the thread. They are available in sizes that measure pitch diameters up to 2 inches. Each micrometer

has a given range of number of threads per inch that can be measured correctly.

- *Ball micrometer.* This type of micrometer (not shown) has a rounded anvil and a flat spindle. It's used to check the wall thickness of cylinders, sleeves, rings, and other parts that have a hole bored in a piece of material. The rounded anvil is placed inside the hole and the spindle is brought into contact with the outside diameter. Ball attachments that fit over the anvil of regular outside micrometers are also available. When using the attachments, you must compensate for the diameter of the ball as you read the micrometer.
- *Blade micrometer.* A blade micrometer (not shown) has an anvil and a spindle that are thin and flat. The spindle does not rotate. This micrometer is especially useful in measuring the depth of narrow grooves, such as an O-ring seat on an outside diameter.
- *Groove micrometer.* A groove micrometer (not shown) looks like an inside micrometer with two flat disks. The distance between the disks increases as you turn the micrometer. It is used to measure the width of grooves or recesses on either the outside or the inside diameter. The width of an internal O-ring groove is an excellent example of a groove micrometer measurement.

Dial indicator

Machinery repair workers use dial indicators to set up work in machines and to check the alignment of machinery. You'll need a lot of practice to become proficient in the use of this instrument. You should use it as often as possible to help you do more accurate work.

Dial indicator sets usually have several components that permit a wide variation of uses. For example, the contact points allow use on different types of surfaces, the universal sleeve permits flexibility of setup, the clamp and holding rods permit setting the indicator to the work, the hole attachment indicates variation or run out of inside surfaces of holes, and the tool post holder can be used in lathe setups and shows some practical applications of dial indicators.

Dial indicators come in different degrees of accuracy. Some will give readings to one ten-thousandth (0.0001) of an inch, while others will indicate to only one five-thousandth (0.005) of an inch. Dial indicators also differ in the total range or amount they will indicate. If a dial indicator has a total of 100 one-thousandths (0.100) of an inch in graduations on its face and has a total range of 200 one-thousandths (0.200) of an inch, the needle will only make two revolutions before it begins to exceed its limit and jams up. The degree of accuracy and range of a dial indicator is usually shown on its face. Before you use a dial indicator, carefully depress the contact point and release it slowly. Rotate the movable dial face so the dial needle is on zero. Depress and release the contact point again, and check to make sure the dial pointer returns to zero; if it does not, have the dial indicator checked for accuracy.

Vernier caliper

You can use a *vernier caliper* (Figure 5.2) to measure both inside and outside dimensions. Position the appropriate sides of the jaws on the surface to be measured, and read the caliper from the side marked inside or outside as required. There is a difference in the zero marks on the two sides that is equal to the thickness of the tips of the two jaws, so be sure to read the correct side. Vernier calipers are available in sizes ranging from 6 inches to 6 feet and are graduated in increments of thousandths (0.001) of an inch. The scales on vernier calipers made by different manufacturers may vary slightly in length or number of divisions; however, they are read basically the same way.

Dial vernier caliper

A *dial vernier caliper* (Figure 5.3) looks much like a standard vernier caliper and is also graduated in thousandths (0.001) of an inch. The main difference is that instead of a double scale, as on the vernier caliper, the dial vernier has the inches marked only along the main body of the caliper and a dial that indicates thousandths (0.001) of an inch. The range of the dial vernier caliper is usually 6 inches.

Dial bore gauge

The *dial bore gauge* is one of the most accurate tools used to measure a cylindrical bore or check a bore for out-of-roundness or taper (Figure 5.4). It does not give a direct measurement; it gives you the amount of deviation from a preset size or the amount of deviation from one part of the bore to another.

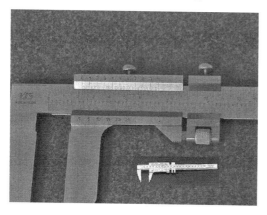

Figure 5.2 Vernier caliper.

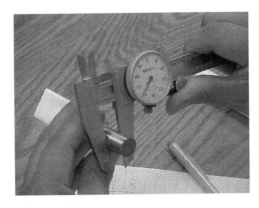

Figure 5.3   Dial vernier caliper.

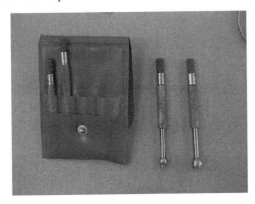

Figure 5.4   Dial bore gauge.

A master ring gauge is used to preset the gauge. A dial bore gauge has two stationary spring-loaded points and an adjustable point to permit a variation in range. These three points are evenly spaced to allow accurate centering of the tool in the bore. A fourth point, the tip of the dial indicator, is located between the two stationary points. By simply rocking the tool in the bore, you can observe the amount of variation on the dial. Most models are accurate to within one ten-thousandth (0.0001) of an inch.

Internal groove gauge

The *internal groove gauge* (not shown) may be used to measure the depth of an O-ring groove or other recesses inside a bore. This tool lets you measure a deeper recess and one located farther back in the bore than you could with an inside caliper. As with the dial bore gauge, you must set this tool with gauge blocks, a vernier caliper, or an outside micrometer. The reading taken from the dial indicator on the groove gauge represents the difference between the desired recess or groove depth and the measured depth.

Figure 5.5 Vernier bevel protractor.

Universal vernier bevel protractor

The *universal vernier bevel protractor* (Figure 5.5) is used to lay out or measure angles on work to very close tolerances. The vernier scale on the tool permits measuring an angle to within 1/12° (five minutes) and can be used completely through 360°. Interpreting the reading on the protractor is similar to the method used on the vernier caliper.

Universal bevel

The *universal bevel* has an offset in the blade. The offset makes it useful for bevel gear work and to check angles on lathe work pieces that cannot be reached with an ordinary bevel. Set and check the universal bevel with the protractor, or another suitable angle-measuring device, to get the angle you need.

Use a gear tooth to measure both the thickness of a gear tooth on the pitch circle and the distance from the top of the tooth to the pitch chord at the same time. Read the vernier scale on this tool in the same way as with other verniers, but note that graduations on the main scale are 0.020 inches apart instead of 0.025 inches.

Cutter clearance gauge

The *cutter clearance gauge* is one of the simplest to use. You can gauge clearance on all styles of plain milling cutters that have more than eight teeth and a diameter range from 1/2 inch to 8 inches. To gauge a tooth with this instrument, bring the surfaces of the *V* into contact with the cutter and lower the gauge blade to the tooth to be gauged. Rotate the cutter sufficiently to bring the tooth face into contact with the gauge blade. If the angle of clearance on the tooth is correct, it will correspond with the angle of the gauge blade. Cutter clearance gauges that have an adjustable gauge blade to check clearance angles of 0° to 30° are also available.

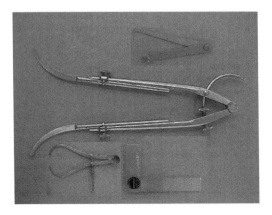

Figure 5.6 Surface gauge.

Inside and outside calipers

Inside and outside calipers, as seen in Figure 5.6, are primarily layout tools to transfer dimensions to drawings, and are used to verify dimensions when used with measurement hand tools, such as dial vernier calipers (Figure 5.3).

Fixed gauges

Fixed gauges cannot be adjusted. Generally, they can be divided into two categories: graduated and nongraduated. The accuracy of your work, when you use fixed gauges, will depend on your ability to compare between the work and the gauge. For example, a skilled machinist can take a dimension accurately to within 0.005 of an inch or less using a common rule. Experience will increase your ability to take accurate measurements.

Graduated gauges

Graduated gauges are direct reading gauges that have scales inscribed on them, enabling you to take a reading while using the gauge. The gauges in this group are rules, scales, thread gauges, center gauges, feeler gauges, and radius gauges.

Rules
- The *steel rule with holder set* (Figure 5.7) is convenient for measuring recesses. It has a long tubular handle with a split chuck for holding the ruled blade. The chuck can be adjusted by a knurled nut at the top of the holder, allowing the rule to be set at various angles. The set has rules ranging from 1/4 inch to 1 inch in length.

Figure 5.7 Steel rule with holder.

- The *adjustable protractor* (Figure 5.8) is used to determine an existing angle or lay out a required angle on material. When I was in the Navy, the protractor was an important part of the damage control tools used to determine the angle of support beams necessary to stop the flooding from a hole in the ship's hull.
- Another useful device is the *keyseat rule,* which has a straightedge and a 6-inch machinist's-type rule arranged to form a right angle square. This rule and straightedge combination, when applied to the surface of a cylindrical work piece, makes an excellent guide for drawing or scribing layout lines parallel to the axis of the work. This device is very convenient when making keyseat layouts on shafts.

You must take care of your rules if you expect them to give accurate measurements. Do not allow them to become battered, covered with rust, or otherwise damaged so that the markings cannot be read easily. Never use rules for scrapers. Once rules lose their sharp edges and square corners, their accuracy is decreased.

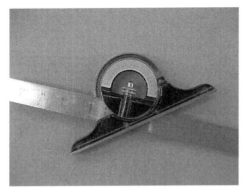

Figure 5.8 Adjustable protractor.

Scales

- A scale is similar in appearance to a rule, since its surface is graduated into regular spaces. The graduations on a scale, however, differ from those on a rule because they are either larger or smaller than the measurements indicated. For example, a half-size scale is graduated so that 1 inch on the scale is equivalent to an actual measurement of 2 inches; a 12-inch-long scale of this type is equivalent to 24 inches. A scale, therefore, gives proportional measurements instead of the actual measurements obtained with a rule. Like rules, scales are made of wood, plastic, or metal, and they generally range from 6 to 24 inches.

Acme thread tool gauge

The *thread gauge* (Figure 5.9) is used both to grind the tool used to machine Acme threads and to set up the tool in the lathe. The sides of the Acme thread have an included angle of 29° (14 1/2° to each side), and this is the angle made into the gauge. The width of the flat on the point of the tool varies according to the number of threads per inch. The gauge provides different slots for you to use as a guide when you grind the tool. It's easy to set up the tool in the lathe. First, make sure that the tool is centered on the work as far as height is concerned. Then, with the gauge edge laid parallel to the centerline of the work, adjust the side of your tool until it fits the angle on the gauge very closely.

Center gauge

Use the *center gauge* (Figure 5.10) like the Acme thread gauge. Each notch and the point of the gauge have an included angle of 60°. Use the gauge primarily to check and to set the angle of the V-sharp and other 60° standard threading tools. You may also use it to check the lathe centers. The edges are graduated into 1/4, 1/24, 1/32, and 1/64 inch for ease in determining the pitch of threads on screws.

Figure 5.9   Thread gauge.

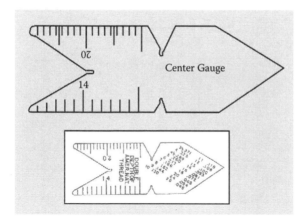

Figure 5.10 Center gauge.

Figure 5.11 Feeler (thickness) gauge.

Feeler gauge

Use a *feeler (thickness) gauge* (Figure 5.11) to determine distances between two closely mating surfaces. When you use a combination of blades to get a desired gauge thickness, try to place the thinner blades between the heavier ones to protect the thinner blades and to prevent them from kinking. Do not force blades into openings that are too small; the blades may bend and kink. To get the feel of using a feeler gauge, practice with it on openings of known dimensions.

Radius gauge

The *radius gauge* (Figure 5.12) is often underrated in its usefulness to the machinist. Whenever possible, the design of most parts includes a radius located at the shoulder formed when a change is made in the diameter. This gives the part an added margin of strength at that particular place. When

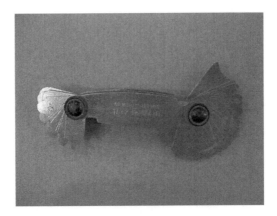

Figure 5.12   Radius gauge.

a square shoulder is machined in a place where a radius should have been, the possibility that the part will fail by bending or cracking is increased. The blades of most radius gauges have both concave (inside curve) and convex (outside curve) radii in the common sizes.

Nongraduated gauges

Nongraduated gauges are used primarily as standards, or to determine the accuracy of form or shape. They include the straightedge, machinist's square, sine bar, parallel block, gauge block, ring gauge and plug gauge, and thread-measuring wire. I'll explain the use of these gauges in the following paragraphs.

- *Straightedges.* Straightedges (Figure 5.13) look very much like rules, except they are not graduated. They are used primarily to check sur-

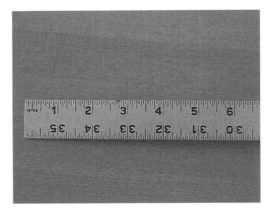

Figure 5.13   Straightedge.

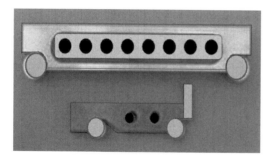

Figure 5.14   Sine bar.

faces for straightness; however, they can also be used as guides to help draw or scribe straight lines. Always keep a straightedge in a box when it is not in use. Some straightedges are marked with two arrows, one near each end, which indicate balance points. When a box is not provided, place resting pads on a flat surface in a storage area where no damage to the straightedge will occur from other tools. Then, place the straightedge so the two balance points sit on the resting pads.

- *Machinist's square.* The most common type of machinist's square has a hardened steel blade securely attached to a beam. This instrument is very useful in checking right angles and in setting up work on shapers, milling machines, and drilling machines. The size of machinist's squares ranges from 1 1/2 to 36 inches in blade length. You should take the same care of machinist's squares, in storage and use, as you do with a micrometer.
- *Sine bar.* A sine bar (Figure 5.14) is a precision tool used to establish angles that require extremely close accuracy. When used in conjunction with a surface plate and gauge blocks, angles are accurate to 1 minute (1/60°). The sine bar is used to measure angles on work and to lay out an angle on work to be machined, or work may be mounted directly to the sine bar for machining. The cylindrical rolls and the parallel bar, which make up the sine bar, are all precision ground and accurately positioned to permit such close measurements. Be sure to repair any scratches, nicks, or other damage before you use the sine bar, and take care in using and storing the sine bar.
- *Parallel blocks.* Parallel blocks (Figure 5.15) are hardened, ground steel bars that are used to lay out work or set up work for machining. The surfaces of the parallel block are all either parallel or perpendicular, as appropriate, and can be used to position work in a variety of setups with accuracy. They generally come in matched pairs and in standard fractional dimensions. Use care in storing and handling them to prevent damage. If it becomes necessary to regrind the parallel blocks, be sure to change the size stamped on the ends of the blocks.

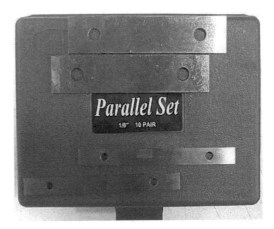

Figure 5.15 Parallel blocks.

- *Gauge blocks.* Gauge blocks (Figure 5.16) are used as master gauges to set and check other gauges and instruments. Their accuracy is from two one-millionths (0.000002) of an inch to eight one-millionths (0.000008) of an inch, depending on the grade of the set. To visualize this minute amount, consider that the average thickness of a human hair divided 1,500 times equals 0.000002 inches. This degree of accuracy applies to the thickness of the gauge block, the parallelism of the sides, and the flatness of the surfaces. To attain this accuracy, a fine grade of harden-

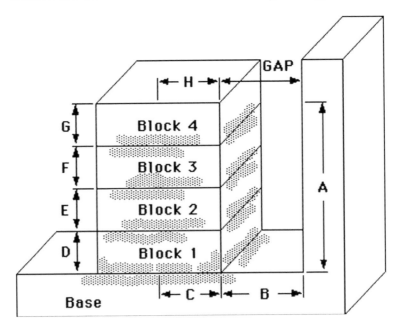

Figure 5.16 Gauge blocks.

able alloy steel is ground and then lapped. The gauge blocks are so smooth and flat that when they are "wrung" or placed one atop the other, you cannot separate them by pulling straight out. A set of gauge blocks has enough different-sized blocks that you can establish any measurement within the accuracy and range of the set. As you might expect, anything so accurate requires exceptional care to prevent damage and to ensure continued accuracy. A dust-free temperature-controlled atmosphere is preferred. After use, wipe each block clean of all marks and fingerprints, and coat it with a thin layer of white petrolatum to prevent rust.

- *Ring and plug gauges.* A *ring gauge* (Figure 5.17) is a cylindrically shaped disk that has a precisely ground bore. Ring gauges are used to check machined diameters by sliding the gauge over the surface. Straight, tapered, and threaded diameters can be checked by using the appropriate gauge. The ring gauge is also used to set other measuring instruments to the basic dimension required for their operation. Normally, ring gauges are available with GO and NOT GO sizes that represent the tolerance allowed for the particular size or job. A *plug gauge* (Figure 5.18) is used for the same types of jobs as a ring gauge. However, it is a solid shaft-shaped bar that has a precisely ground diameter used to check inside diameters or bores.
- *Thread-measuring wires.* These wires provide the most accurate method of measuring the fit or pitch diameter of threads, without going into the expensive and sophisticated optical and comparator equipment. The wires are accurately sized, depending on the number of threads per inch. When they are laid over the threads in a position that allows an outside micrometer to measure the distance between them, the pitch diameter of the threads can be determined. Sets are available that contain all the more common sizes.

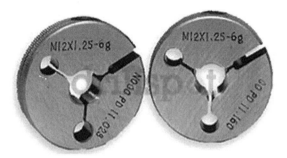

Figure 5.17 Ring gauge.

Figure 5.18   Plug gauge.

The care and maintenance of measuring instruments

The proper care and maintenance of precision instruments are very important to a conscientious machinist.

To help you maintain your instruments in the most accurate and reliable condition possible, the U.S. Navy has established a calibration program that provides calibration technicians with the required standards and procedures, and a schedule of how often an instrument must be calibrated to be reliable.

When an instrument is calibrated, a sticker is affixed to it showing the date the calibration was done and the date the next calibration is due. Whenever possible, you should use the U.S. Navy calibration program to verify the accuracy of your instruments. Some repair jobs, due to their sensitive nature, demand the reliability provided by the program. Information concerning the procedures that you can use in the shop to check the accuracy of an instrument is contained in the following paragraphs.

Micrometers

The micrometer is one of the most used, and often one of the most abused, precision measuring instruments in the shop. Careful observation of the do's and don'ts in the following list will enable you to take proper care of the micrometer you use:

- Always stop the work before taking a measurement. Do *not* measure moving parts because the micrometer may get caught in the rotating work and be severely damaged.
- Always open a micrometer by holding the frame with one hand and turning the knurled sleeve with the other hand. Never open a micrometer by twirling the frame, because such practice will put unnecessary strain on the instrument and cause excessive wear of the threads.
- Apply only moderate force to the knurled thimble when you take a measurement. Always use the friction slip ratchet if there is one on the instrument. Too much pressure on the knurled sleeve will not only result in an inaccurate reading, but also may cause the frame to spring, forcing the measuring surfaces out of line.

- When a micrometer is not in actual use, place it where it is not likely to be dropped. Dropping a micrometer can cause the frame to spring; if dropped, the instrument should be checked for accuracy before any further readings are taken.
- Before a micrometer is returned to stowage, back the spindle away from the anvil; wipe all exterior surfaces with a clean, soft cloth; and coat the surfaces with a light oil. Do not reset the measuring surfaces to close contact because the protecting film of oil on these surfaces will be squeezed out.
- A micrometer caliper should be checked for zero setting (and adjusted when necessary) as a matter of routine to ensure that reliable readings are being obtained. To do this, proceed as follows:
  1. Wipe the measuring faces, making sure that they are perfectly clean, and then bring the spindle into contact with the anvil. Use the same moderate force that you ordinarily use when taking a measurement. The reading should be zero; if it is not, the micrometer needs further checking.
  2. If the reading is more than zero, examine the edges of the measuring faces for burrs. Should burrs be present, remove them with a small slip of oilstone, clean the measuring surfaces again, and then recheck the micrometer for zero setting.
  3. If the reading is less than zero, or if you do not obtain a zero reading after making the correction described in step 2 (above), you will need to adjust the spindle-thimble relationship. The method for setting zero differs considerably between makes of micrometers. Some makes have a thimble cap that locks the thimble to the spindle, some have a special rotatable sleeve on the barrel that can be unlocked, and some have an adjustable anvil.
- To make adjustments to micrometers, follow these steps:
  1. To adjust the *thimble-cap type*, back the spindle away from the anvil, release the thimble cap with the small spanner wrench provided for that purpose, and bring the spindle into contact with the anvil. Hold the spindle firmly with one hand and rotate the thimble to zero with the other; after zero relation has been established, rotate the spindle counterclockwise to open the micrometer, and then tighten the thimble cap. After tightening the cap, check the zero setting again to be sure the thimble-spindle relation was not disturbed while the cap was being tightened.
  2. To adjust the *rotatable sleeve type*, unlock the barrel sleeve with the small spanner wrench provided for that purpose, bring the spindle into contact with the anvil, and rotate the sleeve into alignment with the zero mark on the thimble. After completing the alignment, back the spindle away from the anvil, and retighten the barrel sleeve–locking nut. Recheck for zero setting to be sure you did not disturb the thimble-sleeve relationship while tightening the lock nut.

3. To set zero on the *adjustable anvil type,* bring the thimble to zero reading, lock the spindle if a spindle lock is provided, and loosen the anvil lock screw. After you have loosened the lock screw, bring the anvil into contact with the spindle, making sure the thimble is still set on zero. Tighten the anvil setscrew lock nut slightly, unlock the spindle, and back the spindle away from the anvil; then lock the anvil setscrew firmly. After locking the setscrew, check the micrometer for zero setting to make sure you did not move the anvil out of position while you tightened the setscrew.

- The zero check and methods of adjustment of course apply directly to micrometers that will measure to zero; the *procedure for larger micrometers* is essentially the same except that a standard must be placed between the anvil and the spindle to get a zero measuring reference. For example, a 2-inch micrometer is furnished with a 1-inch standard. To check for zero setting, place the standard between the spindle and the anvil and measure the standard. If zero is not indicated, the micrometer needs adjusting.
- Inside micrometers can be checked for zero setting and adjusted in about the same way as a micrometer caliper; the main difference in the method of testing is that an accurate micrometer caliper is required for transferring readings to and from the standard when an inside micrometer is being checked.
- Micrometers of all types should be disassembled periodically for cleaning and lubrication of internal parts. When this is done, each part should be cleaned in noncorrosive solvent, completely dried, and then given a lubricating coat of watchmaker's oil or similar light oil.

Vernier gauges

Vernier gauges also require careful handling and proper maintenance if they are to remain accurate. The following instructions apply to vernier gauges in general:

- Always loosen a gauge into position. Forcing, besides causing an inaccurate reading, is likely to force the arms out of alignment.
- When taking a measurement, use only gentle pressure on the fine adjustment screw. Heavy pressure will force the two scales out of parallel.
- Before putting a vernier gauge away, wipe it clean and give it a light coating of oil. (Perspiration from hands will cause the instrument to corrode rapidly.)

Dials

Dial indicators and other instruments that have a mechanically operated dial as part of their measurement features are easily damaged by misuse and lack of proper maintenance. The following instructions apply to dials in general:

- As previously mentioned, be sure the dial you have selected to use has the range capability required. When a dial is extended beyond its design limit, some lever, small gear, or rack must give to the pressure. The dial will be rendered useless if this happens.
- Never leave a dial in contact with any surface that is being subjected to a shock (such as hammering a part when dialing it in) or an erratic and uncontrolled movement that could cause the dial to be damaged.
- Protect the dial when it is not being used. Provide a storage area where the dial will not receive accidental blows and where dust, oil, and chips will not contact it.
- When a dial becomes sticky or sluggish in operating, it may be either damaged or dirty. You may find that the pointer is rubbing the dial crystal or that it is bent and rubbing the dial face. Never oil a sluggish dial. Oil will compound the problems. Use a suitable cleaning solvent to remove all dirt and residue.

Chapter six

Measuring and analysis for continuous improvement

It is essential that every company put into place a continuous improvement philosophy throughout the entire organization. The operating infrastructure itself must be designed to incorporate every element necessary to identify all of the opportunities for improvement, and the implementation of improvement projects.

Everyone in the organization must be encouraged to come forward with ideas for improving products, processes, systems, productivity, and the working environment. It is also necessary that improvement suggestions are documented and followed up on in order to keep track of the history of improvement suggestions.

The person or department submitting the improvement suggestion should receive recognition and follow-up on that input. When upper management implements an organization-wide improvement project, the documentation of the project history must be formally captured and represented by a vast amount of data and charts. The problem with this type of project is the cost and the amount of resources it takes to support the project. Additionally, in the majority of these types of larger projects that are dictated by upper management, the longevity is typically short, and it takes resources to initiate many of the improvements that outweigh the value added by the long-term improvement in the company. When I conduct an assessment of an organization, the majority of the proof of continuous improvement is demonstrated by vast amounts of charts and data with no real value.

Determining who should identify improvement opportunities within a company can be difficult. The key is to simulate the onslaught of a virus throughout the organization below the level of information filters (middle management). When this is done successfully, there can be real improvements and a large cost savings for the organization.

For example, if you had a company of 500 employees, what would you rather see?

1. One major change in an organization driven by management that is illustrated through data to save $30,000 in 2008

   or

2. Of the 500 employees, 400 are reducing their nonproductive direct labor by .03 percent

Based on 2,080 work hours per year at an average of $15 an hour for 400 employees, the yearly savings for the second choice would total more than $374,000 in savings and a reduction by .03 percent of wasted time per employee. The cost of recourses to produce the $374,000 yearly savings is usually very minor. In addition, because the workforce, and not senior management, drove the cost savings, the program can be successfully administered for a longer period of time.

Most lean manufacturing techniques are driven from the top of the organization to the bottom of the organization. The changes to production are usually based on eliminating wasted movement and redundancy in functions. The lean manufacturing techniques work well in larger manufacturing organizations where high-quantity runs are performed. This is not to say that lean manufacturing concepts are not effective, but that in using these concepts, many small changes that produce higher savings can be overlooked because they are not quantified or recognized as savings, and there is no highly visible major company drive or array of data and charts that provide an elusion of major savings. Yes, the term *elusion* is used because the data and charts do not reflect the lost hours taken away from direct labor, and the other recourses necessary to justify and produce the so-called savings.

Most of you may be familiar with the common quantified opportunities for improvement that are identified on all the bookshelves, such as ISO 9000 standards, lean manufacturing, Six Sigma, and Gamba Kiezan. The list usually looks like this:

- Data of process and product characteristics, and their trends
- Records of product nonconformities
- Customer satisfaction and dissatisfaction, and other customer feedback
- Market research and analysis of competitive products
- Feedback from employees, suppliers, and other interested parties
- Internal and external audits of the quality system

In addition to the above-listed systems for continuous performance, monitoring and special assessment projects may be initiated to identify opportunities for improvement in other areas. Examples include the following:

- Machine setup and tool changeover times
- Non-value-added use of floor space
- Excessive testing not justified by accumulated results
- Waste of labor and materials
- Excessive cost of nonquality
- Excessive handling and storage

Opportunities for improvement of operations and systems are identified on two levels: continuously, by departmental managers, based on daily feedback from operations and other activities; and periodically, by the management review, based on analysis of long-term data and trends. Most opportunities for the improvement of products are identified primarily by marketing, sales, and engineering.

The evaluation of improvement opportunities is based on daily feedback from operations to the quality assurance department. When appropriate, changes are implemented throughout the system in order to take corrective and preventive actions. Typically, these actions would be triggered by such events as the identification of a nonconforming process or product, a customer complaint, an internal audit finding, and other such specific events.

Opportunities of improvement based on longer term data and trends are evaluated by the management review. They are prioritized with respect to their relevance for reaching the quality policy and quality objectives. When new important opportunities for improvement are not adequately supported by the current policy and objectives, the management review may change the policy and/or establish new quality objectives. This evaluation and prioritizing process is defined in management review meetings.

Opportunities for product improvement are typically evaluated by marketing, engineering, and the top executive management teams.

Implementation of improvement projects requires daily feedback from operations and other activities through the corrective and preventive actions system.

Long-term improvement projects required to fulfill the quality policy, attain quality objectives, or correct unfavorable trends are implemented through special management actions defined by the management review. These actions may be documented in management review minutes, or be issued as directives, memoranda, policy statements, or the like. The corrective and preventive actions system may also be used for this purpose.

Product improvement projects are usually implemented through product briefs or engineering change requests.

As you can see, the above standard practice is appropriate for quantifying continuous improvement, but it is labor intensive. Most companies do not have the staff, or cannot afford to take away staff for the value-added activities needed to justify this method. The majority of today's companies operate with under $5,000,000 in sales per year, and have a workforce of less than 50 employees. In today's market, the ability to correct, adjust, and reduce overhead has created a new type of company that does not build an elaborate system to capture mass amounts of data. Instead, they generate a short memo from the supervisor defining the projected and realized savings. The short memo is more cost-effective and has more validity than consuming the data captured through a full-scale quality system.

Index

A

Acceptance, 7
Acceptance requirements, waivers to, 52
Accuracy
  of dial indicators, 76–77
  of equipment and personnel, 16–20, 20
  expected, 17
  limits as defined on drawings, 16–18
  limits of equipment, 18–20
Acme thread tool gauges, 82
Acronyms, used in drawing design, 63
Adjustable anvil micrometers, adjusting, 90
Adjustable gauges, 74
  ball micrometers, 76
  blade micrometers, 76
  cutter clearance gauges, 79
  depth micrometers, 75
  dial bore gauges, 77–78
  dial indicators, 76–77
  dial vernier calipers, 77
  groove micrometers, 76
  inside and outside calipers, 80
  inside micrometers, 74–75
  internal groove gauges, 78
  micrometers, 74–76
  outside micrometers, 74
  thread micrometers, 75–76
  universal bevels, 79
  universal vernier bevel protractors, 79
  vernier calipers, 77
Adjustable protractors, 81
Aged cast iron, 33
Air hardening steels, 41
Aircraft wings, 60
Allowable minimum, 17
Allowance, 17–18
  relationship to tolerance, 18
Allowed minimum dimension, 17
Alloy steel, 27, 40
Alloys, 25
Aluminum, welding difficulties, 23
Aluminum alloy steels, 42
Amendments, 9–10
American Iron and Steel Institute (AISI), 40
American Society of Mechanical Engineers
  (ASME), 62
American Welding Society (AWS), 36, 38
Appearance, of metals, 25
Appendices, in tailoring checklist, 66

Applied forces, 28
Arc welding, 21, 24
  repairs, 32
ASME Y14.100-2000 practices, 62–63
Associated lists, 65, 66

B

Backfire, 23
Balance points, 85
Ball micrometers, 76
Belt furnaces, 45
Bending stress, in metals, 31
Billets, of steel, 48
Blade micrometers, 76
Bloom, 48
Boeing Aerospace C-17 project, 62
Bottom-up improvements, 94
Brazing, 21, 23, 24, 32
Burring tools, 36

C

Calibration
  basis for requirements, 18
  of equipment, 19
  equipment exempted from, 20
Calibration recall lists, 19
Capacity verification, 9
Car-bottom furnaces, 45
Carbon arc welding, 24
Carbon steel, 27, 40
Carbon-tungsten steels, 41
Carbonitrider furnaces, 45
Care and maintenance, of measuring
     instruments, 87–91
Cast iron, 27
Cast iron cylinder heads, 33
Casting discontinuities, 48–49
Casting mold, 49
Catastrophic failure, 68
Cemetite, 27
Center gauges, 82
Change orders, 9
Charpy V-notch testing, 60
Chemical resistance, 25
Chinese manufacturers, revisions case study,
     xvi–xvii
Chrome steel, 40–41
Chromium hot work steels, 41

Chromium-moly steel, 40
Clearance allowance, 17
Coalescence, 22
Cold shuts, 49–50
Cold welding, 24
Color markings, 25
Combustion, 43
    in oxyacetylene welding, 22
Commander-in-chief, flow of directives, 70
Common metals, 26
Competition, 2
    market research on, 94
Completion dates, 1
Complex stresses, 28
Compression forces, 28, 30
Compression stresses, 28
Computer drive cycle strain machine, 61
Conditions definitions, xv
Configuration items
    defining in contracts, 8
    factors in selection, xvi
    overview, xiv–xvii
Configuration management (CM), xiii
    application of, xiii
    designer as driver of, xiv
    loss through outsourcing, xvi
    overview, xiii–xiv
    turning perception into reality with, 1–2
    using objective evidence in, 3–6
Configuration managers, xiii, 2
Consideration, 7
Continuous belt furnaces, 46
Continuous improvement, measuring and
    analysis for, 93–95
Contract, legal definition, 7
Contract deliverables (CDRLs), 7, 8
Contract review, 5, 8–9
    purpose of, 7
    records of, 10
Contractual requirements, 7–8
    and requests for proposals, 8–10
Control
    final, 51
    first-article, 51
    in-process, 50–51
    receiving, 50
    at verification points, 50
Controlled atmosphere integral quench
    furnaces, 45
Corporate insensitivity, vii
Corrective action reports (CAR), 5
Corrosion, resistance to, 25
Cost-of-poor-quality (COPQ) report, 4, 6
Cost overrides, 1

Cost reduction practices, effects on quality
    assurance, 62
Counteroffer, 7
Cropping the top, 47
Customer satisfaction, 94
Cutter clearance gauges, 79
Cylinder heads, fatigue cracking in, 33–36

D

Decarburization, 43
Deformation, 25, 31
    and strain, 28
Department of Defense (DoD) standards, 62
Depth micrometers, 75
Designers, as drivers of configuration
    management, xiv
Destructive testing, 59
    fracture toughness tests, 60–61
    tensile tests, 59–60
Detrimental impurities, 27
Dial bore gauges, 77–78, 78
Dial indicators, 76–77
Dial vernier calipers, 77
Dials, care and maintenance, 90–91
Diffusion welding, 24
Dimensioning and tolerance, 64
Dip brazing, 24
Direction of lay, 15
Dirt impurities, 27
Discontinuities, 67, 68
    casting, 48–49
    cold shuts, 49–50
    forging, 48
    in metals, 46–48
Documentation, case study, xvii
Downsizing, vii
Drawing design, 62
    acronyms used in, 63
    ASME Y14.100-2000 practices, 62–63
    tailoring checklist, 63
Drawing detail, 64
Drawing format, 64
Drawing media, 64
Drawing notes, 64
Drawing sheet size, 64
Drawing types, 64
Dye penetrant, 53
    inspection, 53–54, 54–55

E

Elastic limit, 31
Elasticity, 25

in metals, 31
Electrically heated salt furnaces, 4
Electricity-heated furnaces, 44
Electromagnetic spectrum chart, 56
Electron beam welding, 25
Electroslag welding, 25
Emergency repairs, material substitutions
    in, 68
End user, role in configuration management,
    xiv
Engine cylinder head repair, 32
Engineering, unsuitability for selecting CIs,
    xvi
Engineering drawings, markings on, 65
Equal and opposite tension stresses, 29
Equipment
    calibration of, 19
    exemption from calibration, 20
    limits and accuracy, 18–20
    nonconforming, 20
    selection of, 18–19
    storage and maintenance, 19
Excessive testing, 94
Explosion, 42
Explosion welding, 24
External burst, 49

F

Fabrication standards
    interpretation of, 66–68
    sea lawyer interpretation, 69–70
Fatigue cracking, 33, 34
    of cylinder heads, 33
    steps in repair, 34–36
Feeler gauges, 83
Ferrous metals, 25, 47
Filler metal, 21
Final control, 51
First-article control, 51
Five whys, 5–6
Fixed gauges, 80
    acme thread tool gauges, 82
    adjustable protractors, 81
    center gauges, 82, 83
    feeler gauges, 83
    gauge blocks, 86–87
    graduated gauges, 80–84
    keyseat rules, 81
    machinist's squares, 85
    nongraduated gauges, 84–87
    parallel blocks, 85, 86
    plug gauges, 87
    radius gauges, 83–84

ring gauges, 87
    rules, 80–81
    scales, 82
    sine bars, 85
    steel rule with holder set, 80, 81
    straightedges, 84–85
    thread gauges, 82
    thread-measuring wires, 87
Flag-note symbols, 62
Flashback, 23
Floor space, non-value-added use of, 94
Flow of directives, 69
Flux cored arc welding, 24
Flux leakage, 53
Forge welding, 24
Forging discontinuities, 48, 49
Fracture toughness tests, 60–61
Free-cutting steel, 41
Friction slip ratchet, 88
Friction welding, 24
Fuel-fired furnaces, 42–43
Full-penetration welds, 57, 58
Function tests, 51
Furnace brazing, 24
Furnace control, 44
Furnace types, 44
    belt furnaces, 45
    car-bottom furnaces, 45
    carbonitrider furnaces, 45
    controlled atmosphere integral quench
        furnace, 45
    high-temperature furnaces, 44
    low-temperature furnaces, 46
    pit furnaces, 46
    pusher furnaces, 45
    rotary furnaces, 46
    shaker furnaces, 45–46
    simple box furnaces, 45
    vacuum furnaces, 45
Furnaces
    electricity-heated, 44
    fuel-fired, 42–43
    furnace control, 44
    in heat treatment, 42
    high-temperature, 44–46
    types of, 44–46

G

Gamba Kiezan, 94
Gamma rays, 56
Gas-fired salt furnaces, 44
Gas metal arc welding (GMAW), 23, 24
Gas tungsten arc welding, 24

Gas welding, 24
Gauge blocks, 86, 87
Gauge clearance, 79
Gauge R&R, 16, 20
Graduated gauges, 80
    acme thread tool gauges, 82
    adjustable protractors, 81
    center gauges, 82, 83
    feeler gauges, 83
    keyseat rules, 81
    radius gauges, 83–84
    rules, 80–81
    scales, 82
    steel rule with holder set, 80, 81
Gray cast iron repair, 32
    and cast iron cylinder heads, 33
    fatigue cracking of cylinder heads, 33–36
    repair steps, 34–36
    stress cracking, 33
Greed
    and lost tribal wisdom, 11
    negative effects on employee growth, vii
Groove micrometers, 76
Group processing, 37

H

Half-size scales, 82
Hand tools. *See also* Measuring instruments
    lost art of, 73–74
Handling and storage, excessive, 94
Hardness, 25
Heat-affected zone (HAZ), 68
    inspection of, 57
Heat treatment, 27
    furnaces used in, 42–44
    materials for, 40–42
    of metals, 38–39
    processes, 39–40
Hierarchy
    in flow of communications, 70
    in product fabrications, 69
High-carbon steel, 27, 41
High-speed steels, 41
High-temperature furnaces, 44–46
Home building analogy, and CM, 1–2
Hot tears, 49, 50
Hot work steels, 41
Human tolerance criteria of accuracy, 20

I

Identification tests, for metals, 25
Improvement opportunities, 93

identification by marketing, sales, and
    engineering, 95
Impurities, in steels, 27
In-process control, 50–51
In-process inspections, 5, 6
Inclusions, nonmetallic, 47
Indefinite terms, in specifications, 63
Induction brazing, 24
Induction welding, 25
Industrial radiography, 56
Infrared brazing, 24
Ingot cropping, 47, 48
Inside calipers, 80
Inside micrometers, 74–75
    checking for zero setting, 90
Instrument care and maintenance, 87
    dials, 90–91
    micrometers, 88–90
    vernier gauges, 90
Intelligence with elasticity, viii
Interference allowance, 18
Interference microscope inspection, 14
Internal burst, 49
Internal groove gauges, 78
Iron, 27
Iron carbide, 27
Iron ulna, 73
ISO 9001:2008, 7
ISO 9000 standards, 94
Isotopes, 56

J

Joint strength, 38

K

Keyseat layouts, 81
Keyseat rules, 81
Knowledge base
    loss of, 11
    removal of, 2

L

Language style, in specifications, 63
Laser beam welding, 25
Lay, 14
    symbols indicating direction of, 15
Lead document, 71
Lean flame, 42
Lean manufacturing
    loss of knowledge base through, 2
    top- *vs.* bottom-driven, 94

Lean Quality Systems, Inc., xi
Liability, and tailoring checklist, 63
Limestone, in steel-making, 47
Limits of accuracy
  allowance, 17–18
  as defined on drawings, 16
  tolerance, 16–17
Load, 27, 30
  and stress, 28
Long-term contracts (LTCs), 7
Long-term improvement projects, 95
Longitudinal sound wave, 58
Lost tribal knowledge, 2
  of project management, 11–12
Low alloy. special purpose steels, 41
Low carbon mold steels, 41
Low-carbon steel, 27
Low-temperature furnaces, 46

M

Machine setup times, 94
Machined diameters, checking, 87
Machinist's squares, 85
Magnetic particle inspection, 33, 52–53
Manufacturing, consequences of
    outsourcing on CM, xvi
Market research, 94
Material substitutions, in emergency repairs,
    68
Measurement identification, 18–19
Measuring, *vs.* gauging, 74
Measuring and test equipment, 19
Measuring instruments, 73–74. *See also* Hand
    tools
  acme thread tool gauges, 82
  adjustable gauges, 74–80
  adjustable protractors, 81
  ball micrometers, 76
  blade micrometers, 76
  care and maintenance, 87–91
  center gauges, 82, 83
  clutter clearance gauges, 79
  depth micrometers, 75
  dial bore gauges, 77–78, 78
  dial indicators, 76–77
  dial vernier calipers, 77, 78
  dials, 90–91
  feeler gauges, 83
  fixed gauges, 80–87
  gauge blocks, 86–87
  graduated gauges, 80–84
  groove micrometers, 76
  inside and outside calipers, 80

  inside micrometers, 74–75
  internal groove gauges, 78
  keyseat rules, 81
  machinist's squares, 85
  micrometers, 74–76, 88–90
  nongraduated gauges, 84–87
  outside micrometers, 74
  parallel blocks, 85, 86
  plug gauges, 87
  radius gauges, 83–84
  ring gauges, 87
  rules, 80–81
  scales, 82
  sine bars, 85
  steel rules, 80
  straightedges, 84–85
  thread gauges, 82
  thread-measuring wires, 87
  thread micrometers, 75–76
  universal bevels, 79
  universal vernier bevel protractors, 79
  Vernier calipers, 77
  Vernier gauges, 90
Melting point, in welding and allied
    processes, 21
Metal, surface characteristics, 12–14
Metal characteristics, 25–26
  bending stress, 31
  elasticity, 31
  iron and steel, 27
  shearing stress, 29
  stress and strain, 27–28
  tension stress, 29
Metallic properties, 25
Metalloids, 25, 26
Metals
  classes of, 25
  heat treatment of, 38–39
  origins of discontinuities in, 46–50
  rate of cooling, 39
Micrometers, 73, 74
  ball micrometers, 76
  blade micrometers, 76
  care and maintenance, 88–90
  consequences of dropping, 89
  depth micrometers, 75
  groove micrometers, 76
  inside micrometers, 74–75
  outside micrometers, 74
  thread micrometers, 75–76
Military standards, 18, 36, 62, 67
  weld symbols in, 38
Missing tribal wisdom, 11
Modulus of elasticity, 31

Moly high speed steels, 41
Moly hot work steels, 41
Muffles, 43
Multisheet drawings, maintenance of, 65

N

NAVSEA 250-1500-standard, 67
Navy Nuclear Program, 62
Navy Subsafe Program, 62
Negative tolerance, 18
Negotiation of terms, 7
Neutral axis, 31
New management enterprise, xiv
    configuration items overview, xiv–xvii
    configuration management overview,
        xiii–xiv
Newton's third law of motion, 28
Non-nuclear NDT, 51
Nonconforming equipment, 20
Nonconforming material report (NCMR),
    4, 5
Nonconforming report, 20
Nondestructive testing, 51–52
    dye penetrant inspection, 53–55
    magnetic particle inspection, 52–53
    radiographic inspection, 55–56
    ultrasonic inspection, 56–58
Nonferrous metals, 25
Nongraduated gauges, 84
    gauge blocks, 86–87
    machinist's squares, 85
    parallel blocks, 85
    plug gauges, 87
    ring gauges, 87
    sine bars, 85
    straightedges, 84–85
    thread-measuring wires, 87
Nonmetallic inclusions, 47, 48
Nonmetals, 25
Nonquality, excessive cost of, 94
NOTES block, 13
Nuclear Subsafe Submarine Navy repair, 69
Nuclear Systems Repair Department, 69
Numbering system, 42
    for heat treatment of metals, 40

O

O-ring grooves, 76, 78
Objective evidence
    and five whys, 5–6
    and quality system planning, 4–5
    using in CM, 3–6

Objectives, 3
Offer, 7
Oil hardening steels, 41, 42
Open-fired furnaces, 43
Order changes, 9–10
Organizational standard requirements,
    70–71
Outside calipers, 80
Outside micrometers, 74, 75
Outsourcing, impact on configuration items,
    xvi
Overhead reduction, 95
Oxide-coated taps, 32
Oxyacetylene welding, 22–23

P

Painting analogy, xiii–xiv
Parallel blocks, 85, 86
Peaks and valleys, 12, 14
Percentage of penetration, 38
Personal growth, knowledge foundation
    for, vii
Personnel, limits and accuracy, 20
Piezoelectric crystals, 57
Piezoelectric transducers, 56
Pig iron, 47
Pit furnaces, 46
Plain carbon steels, 40
Plasma arc welding, 24
Plug gauges, 87, 88
Policy changes, 71
Policy manuals, 4
Policy objectives, 3
Porosity, in steels, 48
Positive tolerance, 18
Pot furnaces, 44
Preventive measures, 5
Prework review, 61–62
Probability descriptions, xvi
Product conformity, 3, 94
Product failure, identifying root causes of,
    xvi–xvii
Product improvement projects, 95
Product quality objectives, 4
Product realization, and prework review, 61
Product requirements, defining in contracts,
    8
Products of combustion, 43
Profilometer, 14
Project angle, 66
Project management
    accuracy of equipment and personnel,
        16–20

destructive testing, 58–61
discontinuities in metals, 46–50
drawing design, perception, and
    execution, 62–66
heat treatment of metals, 38–46
interpretation of fabrication standards
    and specifications, 66–70
lost art of gray cast iron repair, 32–36
lost tribal wisdom of, 11–12
and metal characteristics, 25–31
nondestructive testing, 51–58
organizational standard requirements,
    70–71
prework review, 61–62
surface characteristics of metal, 12–14
and surface finish quality, 14
verification points, 50–51
weld joint design, 36–38
welding an allied processes, 20–25
Purchase orders (P.O.s), 7
Pusher furnaces, 45

Q

Quality management systems, xi
Quality objectives, 3
Quality performance objectives, 3
Quality planning, 3
Quality policy, 3
Quality system audits, 94
Quality system objectives, 4
Quality system planning, 4–5
Quenching, 40

R

Radiant tubes, heating with, 43
Radiographic exposure fundamentals, 57
Radiographic inspection, 55–56
Radius gauges, 83–84
Range capability, for dials, 91
Realization, process of perception to, 1
Receiving control, 50
Recess measurement, 80
Reference documents, 62
Reference measuring and test equipment, 19
Reliability ratings, xvi
Request for proposal (RFP), 8
    and capacity verification, 9–10
    contract review and, 8–9
    contract review records, 10
    and order changes/amendments, 9–10
Resistance brazing, 24
Resistance welding, 24

Responsibility assignment, 7
Retention period, for review of records, 10
Retorts, 43
Review of records, retention period, 10
Revision history, maintenance of, 65
Revisions, errors by sales and marketing,
    xvii
Rich flame, 42
Rickover, Admiral Hyman, 67
Ring gauges, 87
Risk priority numbers, and severity rating,
    xv
Root cause, 5
Rotary furnaces, 46
Rotatable sleeve micrometers, adjusting, 89
Roughness height rating, 12, 13
    maximum permissible, 13
Roughness height values, 14, 16
Roughness width cutoff, 14
Roughness width value, 14
Round-robin methods, 16
Rules, 80–81

S

Safety ratings, xvi
Sales and marketing
    review of contracts by, 8
    unsuitability for CI decisions, xvii, 8
Saturation diving, 22
Scales, 82
Sea lawyer, 69–70
Secretary of the Navy instruction
    (SECNAVINST), 70
Severity rating guide, xiv, xvi
Shaker furnaces, 45–46
Shear wave, 57, 58
Shearing forces, 28, 29
    in rivets, 30
Shearing stress, 28
    in metals, 29
Sheet additions, 65
Sheet deletions, 65
Shielded metal arc welding, 24
Shock resisting steels, 41
Simple box furnaces, 45, 46
Sine bars, 85
Single-direction tolerance, 18
Six Sigma, 94
Skill sets, 1
Slabs, of steel, 48
Slag, 27, 47
Smelting processes, 47
Smoothness, 12

Society of Automotive Engineers (SAE), 40
Soldering, 21, 24
Solid-state welding, 24
Sows, 47
Specifications
    for heat-treating processes, 39
    interpretations of, 66–68
    red flag example, xvii
Stainless steels, 42
Standards, 7
    as basis for calibration requirements, 18
Statements of work (SOWs), 7
Steel, 27
Steel-making process, 47
Steel rule with holder set, 80, 81
Storage and maintenance, of equipment, 19
Straight fluted taps, 32
Straightedges, 84–85
Strain, 28
Stress and strain, 25
    and load, 28
    in metals, 27–28
Stress cracking, 33
Stud welding, 24
Subcontracting, 2
Surface analyzer, 14
Surface characteristics, of metal, 12–14
Surface finish quality, 12
    reading, 14
Surface roughness, 12

T

Tailoring checklist, 63, 64, 65, 66
Tap holes, with pipe threads, 35
Tapping, drill holes for, 34
Tensile forces, 28
Tensile stresses, 28
Tensile tests, 59–60
Tension stress, 30
    in metals, 29
Test equipment, 19
Thermit welding, 25
Thermocouples, 44
Thimble-cap micrometers, adjusting, 89
Thread gauges, 82
Thread-measuring wires, 87
Thread micrometers, 75–76
Titanium alloys, 42
Toaster oven example, xiv
    conditions definitions, xv
    probability descriptions, xvi
    risk priority numbers and severity, xv
    severity, safety, and reliability ratings, xvi

severity rating guide, xiv, xv
Tolerance, 16–17, 64
    relationship to allowance, 18
Tool changeover times, 94
Tool steels, 41
Torch brazing, 24
Touch comparison, 14
Toughness, 25, 59
Tungsten high speed steels, 41
Tungsten hot work steels, 41
Type-S steels, 41

U

Ultrasonic inspection, 56–58
Ultrasonic welding, 24
Undercut
    dangers of, 67–68
    propagation of, 68
Unilateral tolerance, 18
Unit area, 28
Universal bevels, 79
Universal vernier bevel protractors, 79
Unless otherwise specified, 62
U.S. Navy calibration program, 88
USS Dolphin training points, 62–63

V

Vacuum furnaces, 45
Verification points, 50–51
Vernier bevel protractors, 79
Vernier calipers, 77
Vernier gauges, care and maintenance, 90
Very-high-carbon steel, 27
Visual inspection, 58
    diminishment of art, 73
    in nondestructive testing, 51
    of surface finish quality, 14

W

Waste, of labor and materials, 94
Water hardening steels, 41, 42
Water-hardening steels, 41
Waviness height values, 13
Waviness width values, 13
Weight, 25
Weld joint design, 36–38
Weld strength, 38
Weld symbols, 36, 38
Weldability, 25
Welding, 20–21
    oxyacetylene welding, 22–23

Welding processes, 23–25
    primary categories, 24
Work instructions, resolving ambiguity in,
    71
Wrought iron, 27

X

X-rays, 55, 56

Y

Yield point, 31
Yield strength, 31

Z

Zero settings, 90
    of micrometer calipers, 89

Printed and bound by CPI Group (UK) Ltd, Croydon, CR0 4YY

23/10/2024

01777670-0019